CRYSTALLIZATION PROCESSES

This volume comprises the translation of Sections 2 and 3 of *Mechanizm i Kinetika Kristallizatsii (Mechanism and Kinetics of Crystallization)* edited by N. N. Sirota, F. K. Gorskii, and V. M. Varikash. The translation of Sections 1 and 4 is available from Consultants Bureau in a companion volume, under the same editorship, entitled *Solid State Transformations*.

CRYSTALLIZATION PROCESSES

Edited by

N. N. Sirota, F. K. Gorskii, and V. M. Varikash

Institute of Solid State Physics and Semiconductors
Academy of Sciences of the Belorussian SSR, Minsk

Translated from Russian by
Geoffrey D. Archard

Springer Science+Business Media, LLC 1966

Library of Congress Catalog Card Number 66-18734

This Special Research Report is a translation of Sections 2 and 3 of
Mekhanizm i Kinetika Kristallizatsii, edited by N. N. Sirota, F. K. Gorskii,
and V. M. Varikash, published by Nauka i Tekhnika Press in Minsk in 1964
for the Crystal Production Section of the Scientific Council on Solid State
Physics of the Academy of Sciences of the Belorussian SSR, Department
of Solid State Physics and Semiconductors of the Academy of Sciences of
the Belorussian SSR. The Russian text from which the translation was
prepared was thoroughly corrected by the editors.

© 1966 Springer Science+Business Media New York
Originally published by Consultants Bureau in 1966.
Softcover reprint of the hardcover 1st edition 1966

ISBN 978-1-4899-4628-7   ISBN 978-1-4899-4626-3 (eBook)
DOI 10.1007/978-1-4899-4626-3

FOREWORD

Detailed study of the growth of crystals from solutions and melts together with examination of the processes of mass crystallization indicate that, for a quantitative description of the laws governing the generation and growth of crystals, the earlier accepted mechanisms of the development of two- and three-dimensional nuclei are insufficient. Consideration must be given to the amorphization of the growing crystal surfaces, the structure of the original phase, the possibility that the original phase may contain various complexes which participate in the formation of crystals, the presence of dislocations in the developing crystal, and the possibility of transformation without nuclei. Special significance attaches to problems dealing with such questions of the structural relationships of the original and developing phases, concentration fluctuations, and allowance for external actions, such as various fields and stresses. We thus find it necessary to consider a number of specific competing transformation mechanisms.

The need also exists for further development and improvement in the field of phase transformations of the second kind. We can hardly fail to take account of the finer details of phase transformations, with special attention to changes in internal parameters as well as external equilibrium factors. In particular, the theory of phase transformations of the second kind, which is based on the resolution of the thermodynamic potential into a power series of the internal parameter near the Curie temperature, cannot at present be considered altogether satisfactory, not only because of its limited nature and insufficiently founded approach to the analysis of the peculiarities of transformation, but also because it remains impossible to describe by means of this theory the various forms of transformations over a wide range of variations in temperature and pressure.

Among recrystallization processes, a unique position is occupied by phase transformations which take place in solids (polymorphic transformations, transformations of the martensite type, etc.).

This book considers some important problems of crystallization and phase transformations related to certain questions of particular importance in contemporary physics.

The first section of the book is devoted to a consideration of the mechanism and kinetics of crystallization. In a series of articles, relief and the state of the surface of growing crystals are considered theoretically and experimentally, and an attempt is made at an experimental evaluation of the surface energy at the crystal—melt boundary.

At the present time it appears to have become generally accepted that the surface of a growing crystal facet is partly amorphized and rough and possesses complex relief. The first section also presents various views as to the competing mechanisms in crystallization processes, which are compared with experimental data on the temperature-dependence of crystallization parameters, the linear velocity of crystallization, and the growth rate. A change in the crystallization parameters with temperature is associated with the appearance of metastable phases characterized by the presence of complicated S-shaped transformation curves, and by the influence of composition on the mechanism and rate of transformation processes.

Undoubtedly, together with our study of the processes of crystal growth and the formation of new phases we must examine and analyze the processes of dissolution. A number of articles consider the structural and kinetic laws of crystal dissolution.

A most important question in the theory and practice of crystallization is that of the influence of the structure and state of the original phase on the mechanism and kinetics of the formation of the new phase. We should note, however, that as yet insufficient attention has been paid to the connection between the structure of the liquid phase and the mechanism and kinetics of forming the new crystalline phase. We therefore regard favorably the several attempts by the authors in this collection to direct our attention to the role of the structure of liquids in crystallization processes.

The production of single crystals of semiconducting materials is associated with their purification and the controlled distribution of impurities. A number of articles in the second section are devoted to problems of the crystallization of semiconductors.

Despite the great advances in the experimental and theoretical study of phase transformation processes, crystal growth, and mass crystallization, much additional theoretical and systematic experimental investigation is still necessary to fulfill our present national requirements. The development of modern technology has hastened the demand for the production of single crystals with controlled impurities, density, and character of dislocations, and has confronted us with the problem of phase transformations, which to a great extent determine the tempo of technical progress.

A mastering of the techniques of growing single crystals and obtaining materials with the required properties demands a considerable extension of those experimental and theoretical studies having a direct bearing upon the understanding of micro- and macroscopic features in the development of phase transformations. The present volume, which brings together a large amount of experimental and theoretical results, will surely serve to stimulate further study in the interesting and promising field of phase transformations. Most of the articles in this collection have been presented at various All-Union Conferences on the Theory of Crystallization, Thermodynamics, and the Kinetics of Phase Transformations held in the city of Minsk.

N. N. Sirota

CONTENTS

PART ONE

EXPERIMENTAL AND THEORETICAL STUDY OF PROCESSES OF CRYSTALLIZATION

CONTENTS

PART TWO

Effects of External Actions on the Processes of Crystallization

PUBLISHER'S NOTE

The following Soviet journals cited in this book are available in cover-to-cover English translation:

Russian title	English title	Publisher
Akusticheskii zhurnal	Soviet Physics— Acoustics	American Institute of Physics
Fizika metallov i metallovedenie	Physics of Metals and Metallography	Acta Metallurgica
Izvestiya Akademii Nauk SSSR: Otdelenie khimicheskikh nauk	Bulletin of the Academy of Sciences of the USSR: Physical Sciences	Consultants Bureau
Kinetika i kataliz	Kinetics and Catalysis	Consultants Bureau
Kolloidnyi zhurnal	Colloid Journal	Consultants Bureau
Kristallografiya	Soviet Physics— Crystallography	American Institute of Physics
Metallovedenie i termicheskaya obrabotka metallov	Metal Science and Heat Treatment	Consultants Bureau
Tsvetnye metally	The Soviet Journal of Nonferrous Metals	Primary Sources
Uspekhi fizicheskikh nauk	Soviet Physics—Uspekhi	American Institute of Physics
Zavodskaya laboratoriya	Industrial Laboratory	Instrument Society of America
Zhurnal éksperimental'noi i teoreticheskoi fiziki	Soviet Physics—JETP	American Institute of Physics
Zhurnal fizichekoi khimii	Russian Journal of Physical Chemistry	The Chemical Society (London)
Zhurnal neorganicheskoi khimii	Journal of General Chemistry of the USSR	Consultants Bureau
Zhurnal prikladnoi khimii	Journal of Applied Chemistry of the USSR	Consultants Bureau

PART ONE

EXPERIMENTAL AND THEORETICAL STUDY OF PROCESSES OF CRYSTALLIZATION

INTERPHASE SURFACE ENERGY OF SODIUM CHLORIDE AT THE CRYSTAL–MELT BOUNDARY

F. K. Gorskii and A. S. Mikulich

The interphase surface tension at the crystal–melt boundary determines the kinetics of crystallization processes. It is present as a parameter in the formulas of the fluctuation theory of crystallization.

The improvement of the theory for specific applications assumes a knowledge of this parameter. Until recently [1, 2], however, there were no methods of measuring the crystal–melt interphase energy, and the problem was solved in reverse. The interphase energy was found by theoretical formulas from data on the temperature-dependence of the number of centers and the linear crystallization rate [3, 4]. Only a few papers are known in which this quantity is calculated theoretically [5-8].

The subject of the present investigation is a direct experimental determination of the crystal–melt interphase energy, based on the measurement of the contact angle (θ) formed by the melt at the crystal surface.

From the well-known relation of Neumann,

$$\sigma_{sl} = \sigma_{sl} + \sigma_{sl} \cos \theta$$

we may determine the interphase surface energy at the crystal–melt interface (σ_{sl}) if we know the surface energies of the solid (σ_s) and liquid (σ_l) phases.

As a subject for investigation we chose sodium chloride; very reliable theoretical and experimental values are available for its surface energy in the solid state.

In order to compare the values of interphase surface energy obtained at the crystal–melt boundary with data from kinetic experiments, we studied the temperature-dependence of the number of centers and the linear crystallization rate in fused sodium chloride.

Measurement of Contact Angles and Calculation of Interphase Energy

at the Crystal–Melt Boundary

If a drop of liquid is placed on the surface of a solid, the tangent to the surface of separation between the liquid drop and the air at the intersection of all three surfaces forms a definite angle (called the contact angle, θ, which depends on the properties of the three phases) with the solid surface. According to the general theory of flotation processes [9], this angle is determined by the following relation between the free surface energies σ_s, σ_l, and σ_{sl} of the three surfaces of separation:

$$\cos \theta = \frac{\sigma_s - \sigma_{sl}}{\sigma_l} . \tag{1}$$

Normally the contact angle is determined by photographing the drop of liquid lying on the solid surface. This process causes no special difficulties even at high temperatures, if the solid has a higher melting point than the liquid wetting it.

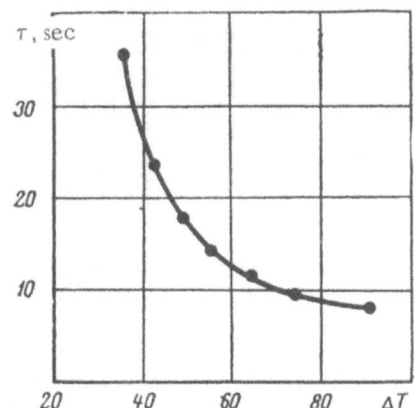

Fig. 1. Contact angle formed by sodium chloride melt with the crystal surface.

Fig. 2. Time for the appearance of the first crystallization center of sodium chloride as a function of supercooling.

In our case it was necessary to measure the contact angle formed at the crystal surface by its own melt, thereby introducing additional experimental difficulties. It proved most advantageous to measure θ from photographs of single crystals with parallel vertical faces immersed in the melt. In this case, as shown in [10], the contact angle is also determined by relation (1).

The temperature of the melt was held close to the melting point. Having been heated to a temperature close to the melting point, the sample was placed in the melt so that its plane surfaces were vertical and then photographed.

Samples of corresponding shape and size were prepared by cutting from a sodium chloride single crystal grown from the melt. The photographs were taken with the help of a telemicroscope TM and microphoto-attachment MFN-3 in transmitted light. Figure 1 clearly shows the curvature of the horizontal surface of the melt and the vertical plane of the crystal immersed in it.

The mean value of the contact angle at the crystal—melt surface for sodium chloride, obtained from not less than twenty measurements, was 28°.

In order to determine σ_{sl} from formula (1), we need to know the surface energy of the solid phase at the melting point. Quite a few papers have been devoted to calculating the surface energy of ionic crystals of the sodium chloride type, but the values obtained for σ_s by various authors differ widely.

For sodium chloride the most reliable values would appear to be those of Frenkel' [11], Zhdanov [12], Zaggeren and Benson [13], and Zadumkin [14].

The question of the value of σ_s for sodium chloride has also not yet been solved experimentally in a definite manner.

In [15] we obtained the value $\sigma_s = 157 \, \text{ergs/cm}^2$ at room temperature. At the melting point this value should be somewhat smaller, since the temperature coefficient $d\sigma/dT$, as shown by measurements for melts, is negative [16].

Assuming that the temperature-dependence of the surface energy is not very different for the solid and the melt, we took the surface energy of sodium chloride at the melting point as $135 \, \text{ergs/cm}^2$. This also corresponds to an approximate estimate for σ_t at the melting point from the formula [17]

$$\frac{\sigma_s - \sigma_l}{\sigma_l} = \frac{Q_m}{Q_v} \, ,$$

(2)

TABLE 1. Temperature-Dependence of the Number of Crystallization Centers
of Sodium Chloride

T, °K	ΔT, °K	$\dfrac{1 \cdot 10^8}{T(\Delta T)^2}$	τ, sec	$\dfrac{1}{\tau}$	I	$\ln I$	$\dfrac{U}{RT}$	$\ln I + \dfrac{U}{RT}$
1041	36	74.5	36	0.0275	24	3.15	4.80	7.95
1034	43	52.3	24	0.0416	36	3.57	4.83	8.40
1028	49	40.5	18	0.0555	48	3.73	4.86	8.69
1022	55	32.3	14	0.0714	62	4.12	4.89	9.01
1013	64	24.1	12	0.0833	73	4.28	4.93	9.21
1003	74	18.2	10	0.1	87	4.46	4.98	9.44
986	91	12.24	8	0.125	109	4.68	5.06	9.74

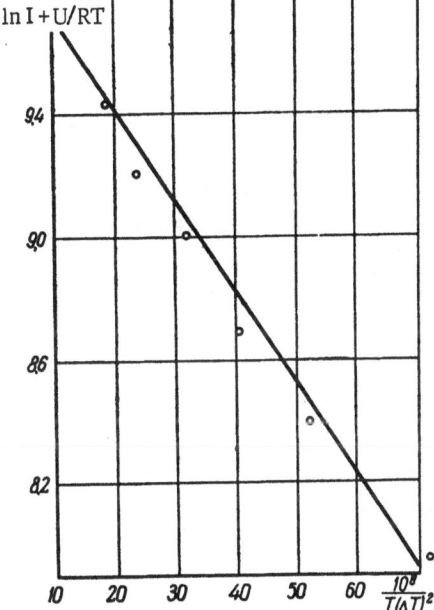

Fig. 3. Variation of $\ln I + U/RT$ with $1/T(\Delta T)^2$ for sodium chloride.

where Q_m is the heat of fusion and Q_v is the heat of vaporization.

By substituting for θ, σ_t, and σ_l [16] in the expression

$$\sigma_{sl} = \sigma_s - \sigma_l \cos \theta,$$

we obtain the value of the interphase surface energy for sodium chloride at the crystal–melt boundary as $\sigma_{sl} = 34$ ergs/cm^2.

Temperature-Dependence of the Number
of Crystallization Centers in Sodium Chloride

According to fluctuation theory, the number of centers (I) arising per unit time in unit volume in a supercooled liquid is determined by the equation

$$I = K \exp\left(-\frac{U}{RT}\right) \exp\left(-\frac{B\sigma^3}{T(\Delta T)^2}\right), \qquad (3)$$

where U is the energy of activation, σ is the surface tension at the nucleus–liquid phase boundary, T is the absolute temperature, ΔT is the supercooling, and K is a coefficient of proportionality.

Comparing the experimental temperature-dependence of the number of crystallization centers (n.c.c.) with the above formula, we can estimate the value of σ. Since, even for slight supercoolings, sodium chloride is characterized by a considerable crystal growth rate, in order to determine the temperature-dependence of the n.c.c. we measured the waiting time for the first center to appear. The waiting time, as we know, may be regarded as inversely proportional to the number of crystallization centers arising in the sample in unit time.

In the experiments we measured the time which elapsed from the moment of immersing the sample in a thermostat at the temperature of the investigation to the moment at which the first center appeared.

The material under investigation was placed in quartz ampoules with an internal diameter of 1.0 to 1.5 mm and a wall thickness of approximately 0.1 mm. The ampoules were sealed. The samples were melted in a bath placed above the thermostat; the temperature of this was above the melting point of sodium chloride (804°C), and could be regulated in order to create different superheatings of the samples.

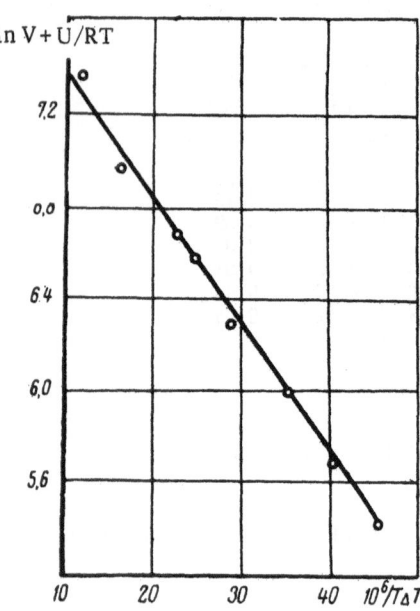

Fig. 4. Variation of ln v+(U/RT) with 1/T∆T for sodium chloride.

The fused sample was placed in the thermostat at the required temperature, steadied beforehand, and automatically maintained. The onset of crystallization was determined in the telemicroscope through a vertical mica window built into the thermostat.

The results of the experiment appear in Table 1, where the waiting time τ is the mean value of not less than twenty observations.

The relation between waiting time τ and supercooling is presented graphically in Fig. 2.

The number of centers I shown in the table was determined from the formula

$$I = \frac{M}{vd} \frac{1}{\tau} ,$$

where the factor M/vd (M is the molecular weight, d is the density, and v is the volume of the sample) is introduced to reduce I to a gram-molecule.

Figure 3 shows the relation between ln I+U/RT and $1/T(\Delta T)^2$. The points fall not too badly on a straight line.

Taking logarithms of formula (3):

$$\ln I = \ln K - \frac{U}{RT} - \frac{B\sigma^3}{T(\Delta T)^2} ,$$

we note that from the slope of this straight line (Fig. 3) we can determine the coefficient of $1/T(\Delta T)^2$, whence, using the formula [18] $B = 16\pi M^2 T_0^2/3kd^2q^2$ (for a spherical nucleus), we can determine the value of the interphase surface energy for the crystal−melt boundary. In the calculations we used the values M = 58.44, d = 2.16 g/cm^3, T_0 = 1077°K, q = 517 J/g.

For the value of the activation energy U we took 10,000 cal/mole, obtained from data [19] on the temperature-dependence of viscosity.

The slope tangent of the straight line in Fig. 3 was $3 \cdot 10^6$, which gives a crystal−melt interphase energy of σ = 29.8 ergs/cm^2.

Temperature-Dependence of the Linear Crystallization Rate of Sodium Chloride

In any consideration of crystal growth processes, the theory of the fluctuational development of two-dimensional nuclei on the growing faces of crystals is generally accepted. The linear crystallization rate (l.c.r.) is proportional to the probability of forming plane nuclei of critical size. The formula for the determination of the l.c.r. has the form

$$v = K_1 \exp\left(-\frac{U}{RT}\right) \exp\left(-\frac{D(\sigma')^2}{T\Delta T}\right) ,$$

(4)

where σ' is the specific peripheral energy.

TABLE 2. Temperature-Dependence of the Linear Crystallization Rate
of Sodium Chloride

T, °K	ΔT, °K	$\dfrac{1}{T\Delta T}$	v, $\dfrac{mm}{min}$	$\ln v$	$\dfrac{U}{RT}$	$\ln v + \dfrac{U}{RT}$
1056	21	$4.5 \cdot 10^{-5}$	2	0.69	4.73	5.42
1053	24	$3.96 \cdot 10^{-5}$	2.7	0.99	4.74	5.73
1048	29	$3.29 \cdot 10^{-5}$	3.3	1.19	4.77	5.96
1043	34	$2.82 \cdot 10^{-5}$	4.6	1.52	4.79	6.31
1038	39	$2.47 \cdot 10^{-5}$	5.9	1.78	4.81	6.59
1034	43	$2.25 \cdot 10^{-5}$	6.5	1.87	4.83	6.70
1013	64	$1.54 \cdot 10^{-5}$	7.8	2.05	4.93	6.98
998	79	$1.26 \cdot 10^{-5}$	11	2.40	4.99	7.39

Taking logarithms of this formula, we obtain

$$\ln v = \ln K_1 - \frac{U}{RT} - \frac{D(\sigma')^2}{T\Delta T},$$

from which we see that the relationship between ln v + U/RT and T/TΔT is linear. Figure 4 shows the experimental values of ln v + U/RT as a function of 1/TΔT. The experimental points fall on a straight line. From the slope of this straight line we can determine σ, or, what is the same thing, σ = σ'/l, where l is the thickness of a monolayer of the nucleus. The l.c.r. was determined by directing the microscope at the moving crystal—melt boundary surface in the sealed quartz ampoules used in the experiments to determine the waiting time for the appearance of the first center. The distance through which the boundary moved was determined with an ocular micrometer. The l.c.r. values obtained for various supercoolings appear in Table 2.

The value of the interphase surface energy for the crystal—melt boundary thus obtained was equal to σ = 23 ergs/cm².

Conclusions

1. The interphase surface energy of sodium chloride at the crystal—melt boundary was determined by measuring the contact angle formed by the melt with the crystal surface and using Neumann's formula.

2. The temperature-dependence of a number of centers was examined together with the linear crystallization rate, and the interphase energy was calculated from formulas derived from fluctuation theory.

3. The agreement between the direct experimental values of interphase surface energy at the crystal—melt boundary with values computed from fluctuation-theory formulas may be regarded as completely satisfactory.

4. The values obtained differ very little from the theoretical estimate of Shcherbakova [7], and also correspond to the minimum value obtained from experimental data [20] on the limiting supercooling of a drop suspended in an inert medium.

Literature Cited

1. F. K. Gorskii, Minsk State Medical Institute: Abstracts of Scientific Session Devoted to the Fortieth Year of the Belorus. SSR (Minsk, 1958).
2. A. Skapski, R. Billups, and D. J. Casavant, J. Chem. Phys. 31(5):1431 (1959).
3. F. K. Gorskii, Zh. Eksp. i Teor. Fiz., 18(1):45 (1948).
4. V. I. Danilov, Zh. Eksp. i Teor. Fiz., 19(3):235 (1949).
5. I. V. Salli, Fiz. Metal. i Metalloved. VIII(5):721 (1959).
6. S. N. Zadumkin, Dokl. Akad. Nauk SSSR 130(4):810 (1960).
7. M. M. Shcherbakov, Koll. Zhurn., XXIII(2):215 (1961).

8.   A. Skapski, Acta Metallurgica, 4:576 (1956).

9.   P. A. Rebinder, Physicochemistry of Flotation Processes [Russian translation] (Metallurgizdat, 1933).

10. J. E. McNutt and G. M. Andes, J. Chem. Phys. 30(5):1300 (1959).

11. Ya. I. Frenkel', Electrical Theory of Solids (Moscow, 1924).

12. V. Zdanov, A. Erschow, and G. Galoshow, Z. Phys. 94:241 (1935).

13. F. Zaggeren and G. Benson, Can. J. Phys. 34(9):985 (1956).

14. S. N. Zadumkin, Izv. Vuz. MVO SSSR, Fizika 2 :151 (1958).

15. A. S. Mikulich, Collection: "Crystallization of Liquids" (Minsk, 1962).

16. Handbook of Chemistry and Physics, Vol. II (1956).

17. B. Ya. Pines, Essays on Metallophysics (Khar'kov, 1961).

18. V. I. Danilov, Collection: "Problems of Physical Metallurgy and the Physics of Metals" (Metallurgizdat, 1949), p. 7.

19. Martens, ed., Physical and Chemical Values [Russian translation], Vol. 10 (1930).

20. E. R. Buckle and A. R. Ubbelohde, Proc. Roy. Soc., A 261:1305 (1961).

RELIEF ON THE SURFACE OF CRYSTALS GROWING FROM SOLUTION

G. R. Bartini,* E. D. Dukova, I. P. Korshunov, and A. A. Chernov

The deposition of material in the process of crystal growth takes place very often by layers. The layers have thicknesses of at least one crystal lattice spacing, i.e., several angstroms. Sources of growth layers on the surface include screw dislocations or plane nuclei arising on the crystal surface as a result of fluctuations of thermodynamic parameters. As a rule, sources of layers are available in fair number, and hence the kinetics of growth and the properties of the developing crystal must depend heavily on the processes of layer deposition and thus on the step structure of the growing surface. For example, the coefficient of impurity capture by a crystal from the melt at high temperatures must in a number of cases rise with increasing thickness of the layers in the crystal. It is further known that inclusions of mother liquor in the crystal are formed with greater probability when the growth involves thicker layers. Hence a study of step relief of the surface is important for the establishment of a connection between growth conditions and the defectiveness of the crystal.

The formation of step structure in crystal surfaces during the process of growth has been considered primarily in theoretical papers [1-3]. Among experimental investigations we may mention that of Howes [4], who studied the distribution of layers from 0.1 μ and upwards in time. The present article contains some results of the first part of the investigation: the effect of the supersaturation of the solution on the relief of a growing face.

The experiments were conducted on crystals of an organic substance, β-methylnaphthalene, grown from alcohol solution.

The crystallites were grown in a special apparatus which ensured steady-state conditions with controllable supersaturation. The apparatus consists of a closed thermostatic system, fused entirely out of molybdenum glass, through which circulates a solution of β-methylnaphthalene (Fig. 1).

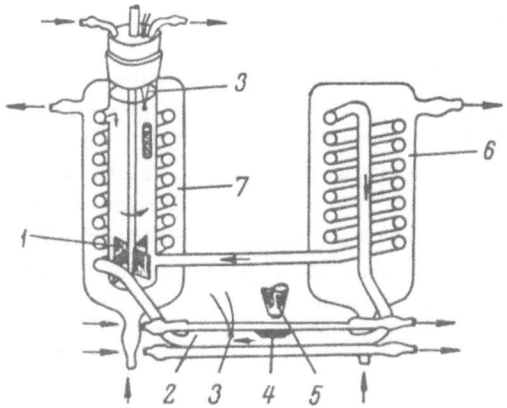

Fig. 1. Scheme of apparatus for growing crystals: 1) propellor; 2, 6, 7) thermostat chambers; 3) thermocouples; 4) sample; 5) microscope tube.

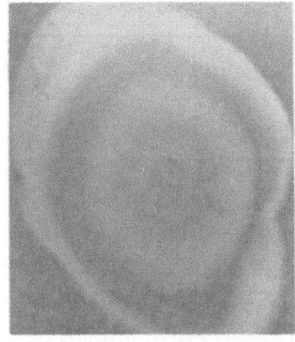

Fig. 2. General view of a growing crystal of β-methylnaphthalene in crossed nicols.

---

*Deceased.

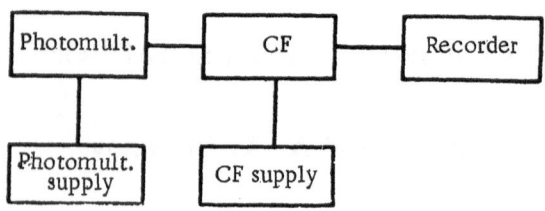

Fig. 3. Block diagram of apparatus for recording
profile diagrams.

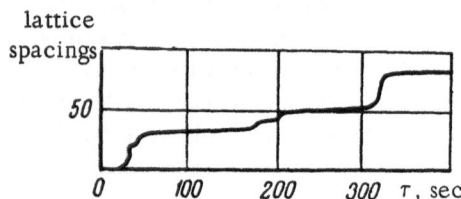

Fig. 4. Profile of crystal surface containing
true macrosteps.

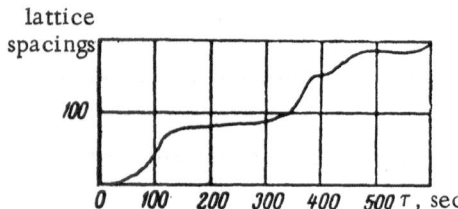

Fig. 5. Profile of crystal surface containing a
kinematic wave of elementary step density.

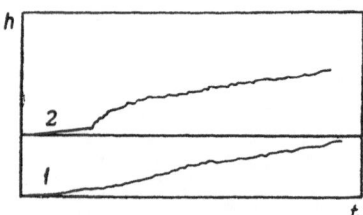

Fig. 6. Formation of a kinematic wave of (1) the
ordinary and (2) the extraordinary type. Curves 1
and 2 are taken at successive instants of time.

The supersaturation of the solution is determined by the difference between the temperature at which saturation of the solution takes place in chamber 7 and that at which crystal growth occurs in chamber 2. The temperature difference was measured by a differential thermocouple calibrated to an accuracy of 0.01°C.

The growth of the crystals was observed in an MIN-4 polarization microscope at a magnification of 200 to 300. For the quantitative study of the crystal surface structure at a fixed point, the field of view was contracted by means of a UF-1 photometer slit. This slit was placed in the plane of the image, which was produced in a photoattachment slipped onto the microscope tube. The breadth of the slit varied from 0 to 1 mm with scale divisions of 10 μ. The experiments were conducted with a slit width of some 100 μ. Thus we actually studied a small part of the crystal surface in the form of a rectangle 0.3 to 0.5 μ wide and 20 to 30 μ long. Insufficient power in the light source prevented further reduction in the size of the region observed.

Crystals of β-methylnaphthalene have large double refraction ($n_g - n_p = 0.27$). If polarized light is passed through such a crystal between crossed nicols, it becomes tinted with interference colors (Fig. 2). The black background obtained with the polarizer and analyzer crossed lights up when even a very thin crystal (10^{-6} cm) appears between them. This may be observed by eye or in the special apparatus, and the crystal thickness thus determined.

In order to determine the intensity of the light which has passed through the crystal, we used a polarization microscope bearing an FEU-19 photoelectron multiplier set immediately behind the UF-1 slit. The supply was a stabilized 1000 V source. The signal from the photomultiplier passed through a cathode follower(CF) on to an EPP-0.9 electronic recorder, which produced a record of the intensity of the light passing through the crystal at the point being examined [5](Fig. 3).

The surfaces of all the growing crystals examined had stepped relief. The very sensitive method described enabled us to reproduce this relief (on a changed scale) on the ribbon of the automatic recorder and examine those of its properties which could not be observed in an ordinary microscopic examination. In those parts of the surface which appear perfectly plane under the microscope, even at high magnifications, the ribbon showed steps of two to four lattice spacings in thickness. These fine steps were at large distances from each other and could be clearly traced. In places the density of the elementary steps increased, and then the recorder ribbon reproduced the shapes of the large steps which could be detected visually. Such steps, tens and even hundreds of lattice spacings high, usually constitute macrosteps. Depending on the density of the elementary steps composing a given macrostep, the shapes take different forms.

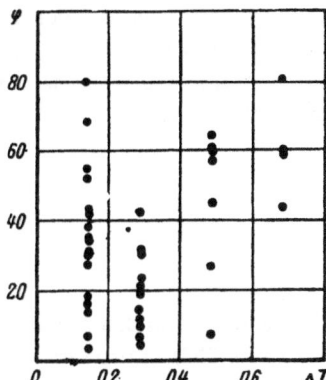

Fig. 7. Slopes of the ends of the macrosteps for various supersaturations.

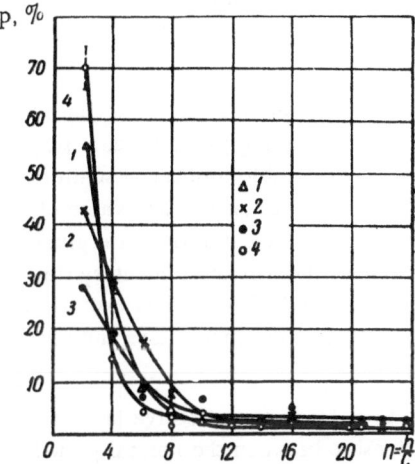

Fig. 8. Height-distribution function of steps
for various supersaturations: 1) $\Delta T = 0.15°C$;
2) 0.1; 3) 0.5; 4) 0.3.

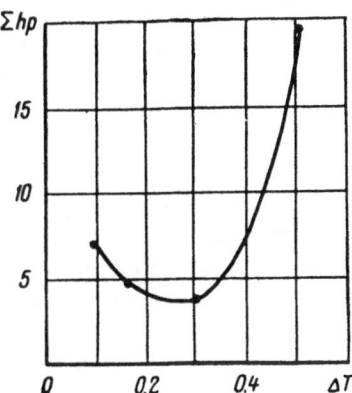

Fig. 9. Mean step height on the surface as a
function of supersaturation.

We observed various types of macroscopic steps.

First, we found "true" macroscopic steps, the end surfaces of which made an angle of more than 40° with the
basal surface of the crystal (Fig. 4).

Secondly, we observed kinematic waves on the surface,
these constituting groupings of elementary steps, the existence
of which was predicted theoretically [1]. Characteristic for
these is the existence of an edge, separating surface sections
of different slope, and a smooth section (Fig. 5). The maximum slope of the surface near the edge in the step shown in
Fig. 5 is 76°. The edge lies to the rear as the echelon moves,
and the spread-out principal bend in the surface lies in front.

In our experiments we also recorded a new form of
kinematic (shock) waves, the form of which was inverse to
that just considered (Fig. 6). These waves move with the
angular point in front and not the spread-out part. We were
able not only to observe the extraordinary waves (their appearance was by no means a rarity) but also to record the
process whereby one of these arose. Figure 6 shows the formation of an extraordinary kinetic wave. It is interesting to
estimate the true slopes of the step profiles from the slope
values obtained on the ribbon, since this angle also enables
us to distinguish the true steps from the kinematic waves.

The slopes of the macrosteps were measured for various
supersaturations. The slopes found are shown in Fig. 7. Despite considerable scatter, the general tendency is obvious.

For small supersaturations ($\Delta T = 0.15°$), different steps
are found, both true and kinematic waves. The greatest number of steps found under these conditions, however, are close
to true ones (slope from 30 to 45°). Then, on increasing the
supersaturation to $\Delta T = 0.3°C$, we have a sharp rise in the
number of macrosteps constituting kinematic waves with a
small slope (from 5 to 25°). On further raising of the supersaturation to 0.65 and 0.7°C, the slope of the macrosteps
again rises.

Thus we can see that for a supersaturation of $\Delta T = 0.3°C$
there is a sharp stratification of the macrosteps and a transformation of true steps into kinematic waves. Then on further
raising of the supersaturation, the density of the elementary
components in the profile of the macrosteps rises, as do the
slopes of their ends. Hence the kinematic waves commonly
transform into steps close to the true type.

We have already noted that steps of various heights,
right up to true macrosteps, are found on the surface of a
crystal. Some of the steps are distributed in disorder, and
some are united into kinematic waves. Within a single
kinematic wave we may also distinguish component steps
of various heights. Apart from the description of surface

roughness (owing to the local slopes present on the surface) just given (Fig. 7), a certain amount of interest may be engendered by a smaller-scale description of the step structure on the faces. One characteristic of such a description is furnished by the density of step distribution with respect to height $\rho(n)$, where ρ is the number of steps intersecting a unit section perpendicular to their fronts. The relative density of the steps

$$p(n) = \frac{\rho(n)}{\sum\limits_{n=1}^{\infty} \rho(n)}$$

constitutes a height distribution function for the steps, normalized to unity. This function was determined experimentally for the surface of crystals grown at various supersaturations (from 0.1 to 0.5°C) in two series of experiments. In each of the series an average of 10 experiments were made for each supersaturation, for solution-saturation temperatures equal respectively to 13.0 and 13.7°C.

For each of the experiments, the relative number of steps of different sizes was reckoned directly from the ribbons of the ÉPP-09 recording the surface profile. The sensitivity of the apparatus only allowed steps with heights of at least two lattice spacings to be identified. Hence the number of elementary steps with n = 1 was determined thus: the sum of the heights of all steps for which n ≥ 2 was subtracted from the total thickness of the crystal and divided by the lattice paramter.

Since the steps and the corresponding jumps on the ribbon of the automatic recorder are sometimes set extremely close together, the main difficulty in deciphering the curves lay in resolving the individual steps; this resolution could not always be made unequivocally. Thus there was a considerable amount of scatter from experiment to experiment in the step densities found, reaching as much as 30%, with an average value of 10%. Nevertheless, general laws were discernible. The relative densities p for various supersaturations are given in Fig. 8. The curves of Fig. 8 indicate first of all the large absolute and relative number of elementary steps on the surface. Their relative density reaches a maximum for a supersaturation of $\Delta T = 0.3$°C. The overall picture of the surface in these conditions is characterized by an intensive disintegration of the macrosteps and kinematic waves, which leads to an increase in the number of low steps.

The increase in the density of steps with supersaturation must lead to a rise in the effective kinetic coefficient characterizing the surface as a whole, i.e., to a nonlinear relationship between the normal growth rate and the supersaturation [6], which is indeed observed.

On passing to large supersaturations ($\Delta T = 0.5$°C), a strong rise in the number of large macrosteps is typical. The inclinations of the ends of these steps are quite large (from 40° upwards). Such steps are also found for supersaturations $\Delta T = 0.1$ and 0.15°C, but in considerably smaller numbers.

The tendencies indicated are clearly seen on the curve relating the mean step height

$$\bar{n} = \sum_{n=1}^{\infty} hp(n)$$

to supersaturation (Fig. 9). This curve has a distinct minimum at $\Delta T = 0.3$°C.

It is known that the formation of defects in crystals in the process of growth is often connected with the existence of macrosteps on the surface. From this point of view the initially obtained results are of especial interest, since these indicate the existence of an optimum supersaturation $\Delta T = 0.3$°C for which the macrosteps play the smallest role. It would be superfluous to add that the complete solution of the problem connecting surface relief with defectiveness requires further serious study.

Literature Cited

1. A. A. Chernov, Usp. Fiz. Nauk 73(2):299-346 (1960).
2. F. C Franc, Growth and Perfection of the Crystals (4):411 (1958).
3. N. Cabrera and D. Vermylea, Growth and Perfection of Crystals, No. 4 (1958).
4. V.R. Howes, Proc. Phys. Soc. 74(5):616-24 (1959).
5. M. I. Kozlovskii and G. G. Lemmlein, Kristallografiya 3:351-57 (1958).
6. A. A. Chernov and E. D. Dukova, Kristallografiya 5(4):655-61 (1960).

MOLECULAR ROUGHNESS OF THE CRYSTAL–MELT BOUNDARY

D. E. Temkin

A number of investigations have been devoted to the analysis of the molecular roughness of the crystalline face [1-3]. In [1], exact results from the two-dimensional Ising model were used to study this question. (The two-dimensional Ising model corresponds to the surface of a crystal in which the atoms can only lie in two levels.) This permitted the establishment of an exact relationship of roughness of the face and the temperature, as well as the determination of the existence of a singular point in this relationship. The degree of roughness (or simply the roughness) is defined as

$$s = (U - U_0)/U_0,$$

(1)

where U_0 is the potential energy of the ideally smooth face (i.e., at $T = 0$) referred to one molecule, and U is the energy of the actual surface.

The singular point on the $s(T)$ curve corresponds to the temperature T_c at which $ds/dT\,|_{T=T_c} = \infty$. Although the degree of roughness s is finite for all temperatures, nevertheless, for $T < T_c$, s is small, and the face can be regarded as practically smooth, while for $T > T_c$ it is rough. For a face with a square lattice (four nearest neighbors), taking into consideration only interactions between nearest neighbors, the temperature T_c is given by the relation

$$\frac{kT_c}{\varepsilon_{11}} \simeq 0.57,$$

(2)

where ε_{11} is the interaction energy between two nearest neighbors in the crystal and k is Boltzmann's constant. For a triangular lattice (six nearest neighbors)

$$\frac{kT_c}{\varepsilon_{11}} \simeq 0.91.$$

(3)

An approximate treatment of the question is given in [2, 3] on the basis of the well-known Bragg–Williams method for the case of a crystal–melt boundary with two possible levels for siting the boundary atoms [2] and for a crystal–vapor boundary with three levels [3]. The use of this method affords an extremely simple analysis of the model of the crystal–melt boundary, without limitations as to the possible number of levels for surface atoms, thereby permitting one to obtain a more complete picture of the form of the phase boundary.

Let us consider the problem of the structure of the crystal–melt boundary in the simplest form. The crystal has the structure of a simple cubic lattice, one of the (100) planes serving as its boundary with the melt. Let us suppose that the liquid has the same density as the crystal and is characterized by the same disposition of atoms (each atom having six nearest neighbors, of which four lie in a plane parallel to the separation boundary). Let us split up the whole crystal–melt system into atomic layers parallel to the (100) plane of the separation boundary, and ascribe a number n to each layer (Fig. 1). In each of these layers there are N atoms, of which in the n-th layer N_n atoms belong to the crystal and the rest ($N - N_n$) to the liquid; the former we call "solid" and the latter "liquid," and the state of the n-th layer we shall characterize by the concentration of "solid" atoms $C_n = N_n/N$.

15

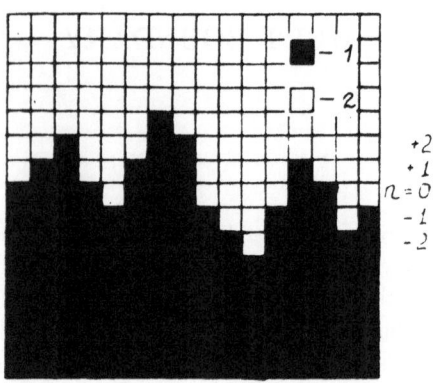

Fig. 1. Schematic representation of phase separation boundary: 1) "solid" atom; 2) "liquid" atom.

Inherent in the nature of this model is the assumption that a certain atom cannot be "solid" unless at least one of its nearest neighbors is also "solid." We shall amplify this assumption somewhat by supposing that a "solid" atom in the $(n+1)$-th layer can be situated only over a "solid" atom of the n-th layer. This excludes from consideration "overhanging" configurations, whose influence on the roughness of the crystal face is yet to be subjected to intensive study. It is clear from what has been said that the quantity C_n must be a falling function of n (as in Fig. 1, we suppose that at $n = -\infty$ we have a solid phase and $C_{-\infty} = 1$, while for $n = +\infty$ we have liquid with $C_{+\infty} = 0$).

The equilibrium values of C_n for a given temperature may be found from the condition for the minimum free energy of the system considered $F(N, C_n, T)$, depending on all the C_n and temperature. It is known that

$$F(N;\ C_n,\ T) = -kT \ln \dot{Z}(N;\ C_n,\ T), \tag{4}$$

where Z is the statistical sum.

In order to describe the statistical sum, let us find the configuration energy of the system for some distribution of "solid" and "liquid" atoms and given values of C_n. As in [1-3], we assume that this energy equals the sum of the energies of the pair interaction of nearest neighbors, and we denote the interaction between two "solid," two "liquid," and "solid" and "liquid" atoms respectively by ε_{11}, ε_{22}, and ε_{12}. The part of the energy of the n-th layer caused by its interaction with the $(n-1)$-th and $(n+1)$-th respectively equals

$$E_{n,\,n-1} = (N/2)\,[C_n\,\varepsilon_{11} + (1 - C_{n-1})\,\varepsilon_{22} + (C_{n-1} - C_n)\,\varepsilon_{12}],$$

$$E_{n,\,n+1} = (N/2)\,[C_{n+1}\,\varepsilon_{11} + (1 - C_n)\,\varepsilon_{22} + (C_n - C_{n+1})\,\varepsilon_{12}]. \tag{5}$$

Here it is considered that the "solid" atom of the $(n+1)$-th layer necessarily lies above a "solid" atom of the n-th layer. The interaction energy of atoms in the n-th layer with one another may be written in the form

$$E_{n,\,n} = N_{11,\,n}\,\varepsilon_{11} + N_{22,\,n}\,\varepsilon_{22} + N_{12,\,n}\,\varepsilon_{12}, \tag{6}$$

where $N_{11,\,n}$, $N_{22,\,n}$, and $N_{12,\,n}$ are respectively the numbers of "solid—solid," "liquid—liquid," and "solid—liquid" bonds for some disposition of atoms in the n-th layer. Since these three quantities satisfy the two relations

$$N_{11,\,n} = 2NC_n - N_{12,\,n}/2,$$

$$N_{22,\,n} = 2N\,(1 - C_n) - N_{12,\,n}/2,$$

expression (6) may be written in the form

$$E_{n,\,n} = 2NC_n\,\varepsilon_{11} + 2N\,(1 - C_n)\varepsilon_{22} + N_{12,\,n}\,w. \tag{7}$$

Here

$$w = \varepsilon_{12} - (\varepsilon_{11} + \varepsilon_{22})/2. \tag{8}$$

The energy of the n-th layer equals $E_{n,\,n} + E_{n,\,n-1} + E_{n,\,n+1}$, and the configuration energy of the system in a certain state characterized by the set of numbers $N_{12,\,n}$ is

$$E(N;\,C_n,\,N_{12,\,n}) = \lim_{m,\,k\to\infty} \sum_{n=-k}^{+m} (E_{n,\,n} + E_{n,\,n-1} + E_{n,\,n+1}) =$$

$$= N\left\{ w + 3\sum_{n=-\infty}^{+\infty} [\varepsilon_{22} + (\varepsilon_{11} - \varepsilon_{22})C_n] \right\} + w\sum_{n=-\infty}^{+\infty} N_{12,\,n}.$$

$$(9)$$

Here k and m tend to infinity in such a way that $C_{-k} = 1$ and $C_{+m} = 0$. Since the dimensions of the system considered are infinite, its energy is also infinite. In what follows, however, we shall be interested in the change in the state of the system connected with a transformation from the ideally smooth surface of separation to a rough one. If we use E_0 to denote the energy of the system in the presence of a smooth boundary set in the zero layer (for this, $C_n = 1$ for $n \le 0$ and $C_n = 0$ for $n \ge 1$), then the change in energy for such a transformation equals

$$\Delta E_s \equiv E(N;\,C_n,\,N_{12,\,n}) - E_0 = w\sum_{n=-\infty}^{+\infty} N_{12,\,n} - 3(\varepsilon_{11} - \varepsilon_{22})\,N\left[\sum_{n=-\infty}^{0}(1 - C_n) - \sum_{n=1}^{\infty} C_n \right].$$

$$(10)$$

Using relation (9), we can write the statistical sum in the form

$$Z(N;\,C_n,\,T) = (j_1)^{-N\sum_{-\infty}^{+\infty}C_n}\,(j_2)^{-N\sum_{-\infty}^{+\infty}(1-C_n)}\,e^{-\frac{N}{kT}\left\{ w+3\sum_{-\infty}^{+\infty}[\varepsilon_{22}+(\varepsilon_{11}-\varepsilon_{22})C_n] \right\}}\sum_{\text{for } N_{12,\,n}} g(N;\,C_n,\,N_{12,\,n})\,e^{-\frac{w}{kT}\sum_{n=-\infty}^{+\infty}N_{12,n}}.$$

$$(11)$$

Here $j_1(T)$ and $j_2(T)$ are the nonconfiguration part of the statistical sum for each "solid" and "liquid" atom respectively (it is assumed that j_1 and j_2 can be separated out from the statistical sum) [4]; $g(N, C_n, N_{12,\,n})$ is the number of states with the given set of numbers $N_{12,\,n}$ for fixed N and C_n. The summation in (11) is carried out over all values of $N_{12,\,n}$ permissible for given N and C_n. If we consider the association between the chemical potentials of the solid (μ_1) and liquid (μ_2) phases and the quantities j_1 and j_2:

$$\mu_1 = 3\varepsilon_{11} - kT\ln j_1,$$
$$\mu_2 = 3\varepsilon_{22} - kT\ln j_2,$$

and call Z_0 the statistical sum for the case of the smooth boundary, then expression (11) takes the form

$$Z(N;\,C_n,\,T) = Z_0 Z_k \exp\left(-N\Delta\mu\left[\sum_{-\infty}^{0}(1 - C_n) - \sum_{1}^{\infty} C_n \right] \Big/ kT \right)$$

$$(12)$$

Here

$$\Delta\mu = \mu_2 - \mu_1 \text{ and } Z_k = \sum_{N_{12,\,n}} g\exp\left(-\frac{w}{kT}\sum_{-\infty}^{+\infty}N_{12,\,n} \right)$$

$$(13)$$

is the configuration statistical sum.

The change in the free energy of the system associated with the formation of a rough surface of separation instead of a smooth one equals

$$\Delta F_s = - kT \ln \frac{Z}{Z_0} = N\Delta\mu \left[\sum_{-\infty}^{0} (1 - C_n) - \sum_{1}^{\infty} C_n \right] - kT \ln Z_k .$$

$$(14)$$

The main difficulty in calculating the statistical sum Z_k lies in finding the function $g(N, C_n, N_{12, n})$. These difficulties have so far not been overcome. The Bragg—Williams method permits an approximate calculation of the statistical sum Z_k; it is calculated as follows. To all configurations with given N and C_n we assign the same mean values $N_{12, n}$:

$$Z_k = \sum_{N_{12, n}} g \exp \left(- \frac{w}{kT} \sum_{n = -\infty}^{+\infty} N_{12, n} \right) \approx$$

$$\approx \exp \left(- \frac{w}{kT} \sum_{n = -\infty}^{+\infty} \overline{N}_{12, n} \right) \sum_{N_{12, n}} g.$$

$$(15)$$

For the value of $\overline{N}_{12, n}$ we choose the mean number of "solid—liquid" pairs for chaotic distribution of NC_n "solid" atoms among N places. In the case considered,

$$\overline{N}_{12, n} = 4NC_n (1 - C_n).$$

$$(16)$$

It is easy to show that

$$\sum_{N_{12, n}} g (N; C_n, N_{12, n}) = N! \left/ \prod_{n = -\infty}^{+\infty} [N (C_n - C_{n+1})] ! \right. .$$

$$(17)$$

In fact, the number of permutations of NC_{n+1} "solid" atoms of the (n+1)-th layer above NC_n "solid" atoms of the n-th layer equals

$$\frac{(NC_n) !}{(NC_{n+1}) ! [N (C_n - C_{n+1})] !} ,$$

and

$$\sum_{N_{12, n}} g = \prod_{n = -\infty}^{+\infty} \frac{(NC_n) !}{(NC_{n+1}) ! [N (C_n - C_{n+1})] !} ,$$

which coincides with expression (17).

After substituting relations (16) and (17) in (15) and then in (14), we have

$$\frac{\Delta F_s}{NkT} = \beta \left[\sum_{-\infty}^{0} (1 - C_n) - \sum_{1}^{\infty} C_n \right] + \gamma \sum_{n = -\infty}^{\infty} C_n (1 - C_n) + \sum_{n = -\infty}^{\infty} (C_{n-1} - C_n) \ln (C_{n-1} - C_n) .$$

$$(18)$$

Here $\beta = \Delta\mu/kT$ and $\gamma = 4w/kT$. In passing from expression (14) to (18), we used Stirling's formula ($\ln N! \simeq N \ln N - N$).

The equilibrium (for a certain temperature) values of concentrations C_n must correspond to the minimum free energy of the system. Differentiating relation (18) with respect to C_n and equating the derivative to zero, we find an equation for determining the equilibrium C_n:

$$\frac{C_n - C_{n+1}}{C_{n-1} - C_n} e^{-2\gamma C_n} = e^{-\gamma + \beta}. \tag{19}$$

The solution of this difference equation of the second order must satisfy two boundary conditions:

$$C_{-\infty} = 1, \quad C_{+\infty} = 0. \tag{20}$$

Moreover, the solution of Eq. (19), in which the minimum of function (18) is realized, should correspond to the positive sign of the quadratic form

$$Q = \sum_{i,j} \frac{\partial^2 (\Delta F_s/NkT)}{\partial C_i \partial C_j} dC_i dC_j > 0. \tag{21}$$

Here the derivatives are taken for values of C_n, forming a solution of Eq. (19):

$$\frac{\partial^2 (\Delta F_s/NkT)}{\partial C_i \partial C_j} = \begin{cases} -2\gamma + \dfrac{1}{C_i - C_{i+1}} + \dfrac{1}{C_{i-1} - C_i} & \text{for } j = i \\[2mm] -\dfrac{1}{C_i - C_{i+1}} & \text{for } j = i+1 \\[2mm] 0 & \text{for } j > i+1. \end{cases}$$

Below we present the results of a numerical solution of Eq. (19). Let us introduce a new variable,

$$z_n = 2\gamma C_n - \gamma + \beta. \tag{22}$$

Then Eq. (19) and boundary conditions (20) take the form

$$\frac{z_n - z_{n+1}}{z_{n-1} - z_n} = e^{z_n}, \quad z_{-\infty} = \gamma + \beta, \quad z_{+\infty} = -\gamma + \beta. \tag{23}$$

In analyzing this equation it is convenient to solve the inverse problem: given two successive values z_0 and z_1, we can find the remaining z_n, $z_{+\infty} = \lim_{n \to \infty} z_n$ and $z_{-\infty}$, and from the values of $z_{-\infty}$ and $z_{+\infty}$ thus found we may determine the values of γ and β corresponding to the given pair of numbers z_0 and z_1. For convenience of calculation, it is useful to write Eq. (23) in the form

$$z_{n+1} = z_n - (z_0 - z_1) e^{\sum\limits_{k=1}^{n} z_k}, \quad z_{-(n+1)} = z_{-n} + (z_0 - z_1) e^{-\sum\limits_{k=0}^{n} z_{-k}}.$$

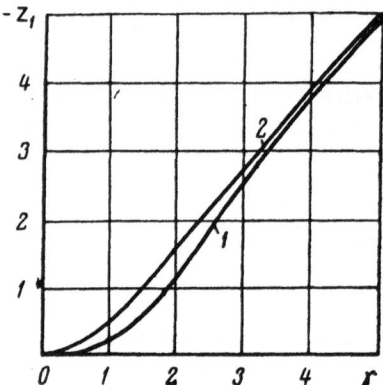

Fig. 2. Connection between the values of z_1 and parameter γ corresponding to the two solutions of equations (23) for $\beta = 0$: 1) solution I; 2) solution II.

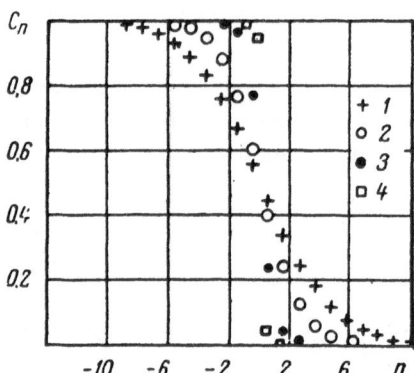

Fig. 3. Concentration distribution of "solid" atoms C_n producing a minimum of the free energy of the system for certain values of parameter $\gamma \equiv 4w/kT_m$: 1) $\gamma = 0.446$; 2) 0.769; 3) 1.889; 4) 3.310.

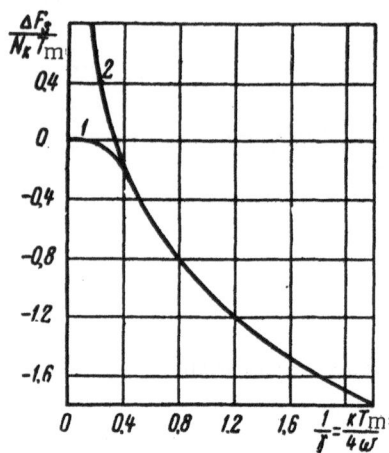

Fig. 4. Changes in free energy of the system on formation of a rough separation boundary instead of a smooth one corresponding to the two solutions of Eq. (23): 1) solution I; 2) solution II.

We shall be interested in positive values of the parameter γ, since these values correspond to the fact that the surface energy of the crystal—melt boundary is positive. (In the case considered, the surface energy of the (100) face of a simple cubic lattice, referred to one atom, is w.)

I. Examination of the Problem when the Temperature is That of the Melting Point T_m

For $T = T_m$, $\beta \equiv \Delta\mu/kT_m = 0$ and $\gamma = 4w/kT_m$. It is not difficult to see that in this case z_n, being an odd function of n, satisfies Eq. (23) and the boundary conditions. For this, two types of solution are possible:

$$z_0 = -z_1, \quad z_{-1} = -z_2 \text{ etc.}$$

$$(\text{or } C_0 = 1 - C_1, \quad C_{-1} = 1 - C_2 \quad \text{etc.}),$$

$$(24)$$

$$z_0 = 0, \quad z_{-1} = -z_1 \text{ etc.}$$

$$(\text{or } C_0 = 1/2, \quad C_{-1} = 1 - C_1 \text{ etc.}).$$

$$(25)$$

Numerical analysis showed that there were no solutions other than those indicated. The quadratic form (21) corresponding to solution (25) is positive, but that corresponding to solution (24) is not sign-determinate. Hence the free energy minimum corresponds to a solution of the (24) type (solution I), since solution II (of the (25) type) corresponds to a saddle point on the $F(C_n)$ surface. The curves of Fig. 2 reflect the relation between z_1 and parameter γ corresponding to each solution. In Fig. 3 by way of example we give solution I for certain values of parameter γ. The change in the free energy of the system associated with the formation of a rough surface of separation instead of a smooth one, and also the roughness s (solution I), are shown as functions of $1/\gamma \equiv kT_m/4w$ in Figs. 4 and 5. The quantity $\Delta F_s/NkT_m$ is calculated from formula (18) ($\beta = 0$) and the roughness s for solution I, according to (10) and (16), from the formula

$$s \equiv \frac{\Delta F_s}{Nw} = 4 \sum_{n=-\infty}^{+\infty} C_n (1 - C_n)$$

(26)

(With $\beta = 0$ for solution I,

$$C_{-n} = 1 - C_{n+1} \quad \text{and} \quad \sum_{-\infty}^{0} (1 - C_n) - \sum_{1}^{\infty} C_n = 0).$$

From the data shown in Figs. 3-5 we see that the state of the separation boundary depends very strongly on the value of parameter γ. With an increase in γ, the width of the boundary falls rapidly, so that its shape tends to become smooth, and $\Delta F_s/NkT_m$ and s fall and tend to zero as $\gamma \to \infty$. For $\gamma \gg 1$: $(-z_1) \sim \gamma$; $C_n \sim e^{-n\gamma}$ ($n \geq 1$, solution I) and $C_n \sim \frac{1}{2} e^{-n\gamma}$ ($n \geq 1$, solution II); $\Delta F_s/NkT_m \sim -2e^{-\gamma}$ (solution I) and $\Delta F_s/NkT_m \sim \gamma/4 - \ln 2$ (solution II); $s \sim 8e^{-\gamma}$. Thus for large γ the width of the surface of separation tends to zero as $1/\gamma$.

In contrast with the results obtained when solving the problem on two levels [1, 2], in the case considered there are no singular points on the curves relating s to $1/\gamma$. According to an exact consideration of the problem on two levels [1], a singular point occurs at $kT_m/2w \approx 0.57$ [in formula (2), ε_{11} must be replaced by $2w$], and in the Bragg—Williams approximation at $kT_m/2w = 1$ [2]. Just as in the problem on three and five levels [1, 3] the curve $s = s(1/\gamma)$ has a point of inflection which may be treated as a point of transition from the state with a smooth boundary to the state with a rough one. The approximate position of the inflection point is $kT_m/2w \approx 0.82$.

II. Examination of the Problem when the System is Supercooled ($T < T_m$)

The main result of the numerical analysis of the equation and boundary conditions (23) reduces to the following. The region of positive values of parameters $\gamma \equiv 4w/kT$ and $\beta \equiv \Delta\mu/kT$ is split by a certain curve into two subregions: in one of these Eq. (23) has no solutions (Fig. 6, region B), and in the other it has two (region A), and one on the boundary. Figure 6 shows on a semilogarithmic scale the boundary between the regions established by a numerical solution of Eq. (23). In region A, just as when there is no supercooling, one of the solutions gives a minimum free energy of the system and the other corresponds to a saddle-point. On approaching the boundary between the regions, the two solutions come closer together, and coincide at the actual boundary.

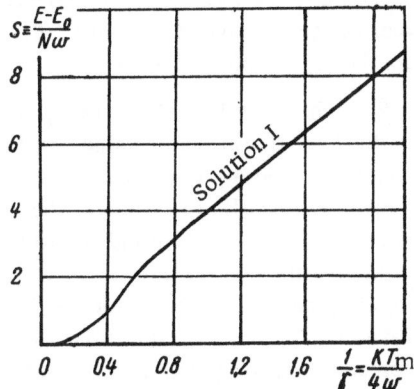

Fig. 5. Roughness as a function of parameter $1/\gamma \equiv kT_m/4w$ (there is an inflection in the curve at $1/\gamma \approx 0.41$).

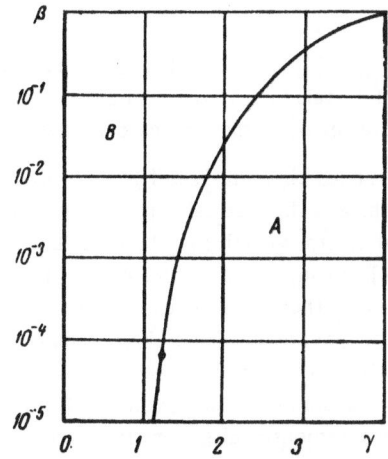

Fig. 6. Curve dividing the region of positive values of parameters β and γ into two subregions.

In region B, which is characterized by greater supercooling than region A, there is no solution to Eq. (23). This means that, in this region of β and γ values, solid and liquid phases separated by a phase boundary cannot coexist. In the case when the system is supercooled, a single-phase state of the system (solid phase) corresponds to the deepest minimum of free energy. In region A, however, on the free energy surface $F(C_n)$ is a series of relative minima corresponding to different positions of the phase boundary. These minimum points are separated from one another by sections with free energy values higher than those at the minima. In this case displacement of the phase boundary requires a certain activation process, and the state in which solid and liquid phases coexist becomes metastable. In region B there are no such metastable states. The phase boundary, having developed, moves without overcoming activation barriers (except those overcome by individual atoms on passing from one phase to the other) until the liquid phase completely disappears. Thus on reaching a certain supercooling the process of transforming the liquid phase into solid takes place "barrierlessly." An analogous result was reached by Roitburd [5] and Cahn [6] on the basis of a phenomenological description of the phase-separation boundary.

As seen from Fig. 6, the supercoolings, or values of parameter $\beta \equiv \Delta\mu/kT$ necessary for the realization of a "barrierless" transformation, depend very strongly on the value of parameter γ, which is determined by the "energy of mixing" $w \equiv \varepsilon_{12} - (\varepsilon_{11} + \varepsilon_{22})/2$. Determination of w for various substances is extremely important and requires detailed analysis, since this parameter determines the quantitative aspect of the problem of phase-boundary structure. For purposes of a qualitative study, let us suppose that the interaction energy ε_{12} of a "solid" atom with a "liquid" one equals the interaction energy for two "liquids" ε_{22} (an analogous assumption was tacitly made in [2]). Then $w = (\varepsilon_{22} - \varepsilon_{11})/2$ and $\gamma = 2(\varepsilon_{22} - \varepsilon_{11})/kT$. If we neglect smallish volume changes as the solid phase passes into the liquid, then for the simple cubic-lattice model considered the heat of fusion L, reduced to one atom, equals $L = 3(\varepsilon_{22} - \varepsilon_{11})$. Hence, assuming, $\varepsilon_{12} = \varepsilon_{22}$, we have $\gamma = (2/3)(L/kT)$.

For the majority of metals $L/kT_m = 0.9\text{-}1.2$ and $\gamma = 2/3 \times L/kT_m = 0.6\text{-}0.8$, while for organic compounds $L/kT_m \gg 1$ (for example, for salol and glycerine $L/kT_m \sim 7.5$ and $\gamma \sim 5$). In the curve of Fig. 5 the inflection point corresponds to $\gamma \sim 2.5$. Hence metals are situated beyond the inflection point on the curve of Fig. 5 and are characterized by high roughness values of the crystal−melt boundary. Such substances as salol and glycerine lie before the inflection point, and the surface of separation in this case is practically smooth.

Let us estimate the amount of supercooling necessary for effecting a "barrierless" transformation. For small supercoolings

$$\beta \equiv \frac{\Delta\mu}{kT} \simeq \frac{L}{kT_m}\frac{\Delta T}{T_m},$$

where $\Delta T = T_m - T$. We see from Fig. 6 that the value of β corresponding to the boundary of regions A and B in the case of metals ($\gamma < 1$) is less than 10^{-5}, and for salol and glycerine ($\gamma \sim 5$) larger than unity. Taking the above estimate for β, in the first case we have $\Delta T/T_m < 10^{-5}$ and in the second $\Delta T/T_m > 0.1$.

In substances characterized by high values of parameter γ ($\gamma > 2.5$), the crystal−melt boundary is practically smooth, and the critical supercoolings corresponding to a "barrierless" transition are large. In this case the process of crystal growth may be described by the well-known theories of spiral growth and two-dimensional generation [2]. In substances with small γ ($\gamma < 2.5$), the surface of separation is diffuse; this eases the crystal growth process for these substances, and the critical supercoolings are small. For supercoolings smaller than critical, the displacement of the phase boundary is accompanied by passage through the activation barrier. In order to describe growth in this case we require a modification of existing theories, allowing for the diffuseness of the phase boundary. For larger than critical supercoolings, the development of the crystal growth process does not require two-dimensional generation, screw dislocations, or any other sources of growth points. In this case, the kinetics of the growth process are determined by the rate of individual acts of transition from one phase into the other.

Literature Cited

1. W. Burton, N. Cabrera, and F. Frank, Collection: "Elementary Crystal Growth Processes" [Russian translation] (IL, 1959).
2. K. Jackson, Collection: "Liquid Metals and Their Solidification" [Russian translation] (Metallurgizdat, 1962).
3. W. W. Mullins, Acta Metallurgica 7(11):746 (1959).
4. T. Hill, Statistical Mechanics [Russian translation] (IL, 1960).
5. A. L. Roitburd, Kristallografiya 7(2):291-99 (1962).
6. J. W. Cahn, Acta Metallurgica 8(8):554-62 (1960).

MECHANISM OF THE GROWTH OF SALOL CRYSTALS FROM THE MELT

D. E. Ovsienko and G. A. Alfintsev

The study of the mechanism of crystal growth is of great significance for the development of the theory of phase transformations and of the means of producing crystals with assigned properties. In practice, crystals are often grown from the melt. The growth mechanism in this case, however, is less clear than for growth from the vapor or from solution. As we know, there exist theories for the growth of perfect and imperfect crystals. The growth of perfect crystals, according to [1-4], takes place by the formation of two-dimensional nuclei. In accordance with this, the relationship between growth rate and supercooling (ΔT) takes the form [2]

$$V = K_1 e^{-\frac{K_2}{T}} e^{-\frac{K_3}{T(\Delta T)}} \tag{1}$$

or [4]

$$V = \frac{K_1}{(\Delta T)^2} e^{-\frac{K_2}{T}} e^{-\frac{K_3}{T(\Delta T)}} , \tag{2}$$

where T is the absolute temperature, and K_1, K_2, and K_3 are constants. The constant K_2 in (1) is associated with the activation energy of the passage of a molecule from the liquid into the crystal, and in (2) K_2 takes account of the energy threshold for uniting whole rows of molecules to the periphery of a two-dimensional nucleus. The constant $K_3 = \pi \rho^2 S T_0 / kQ$ (for a nucleus in the form of a circle), where ρ is the boundary tension at the edge of the two-dimensional nucleus, and the remaining notation is as generally accepted.

In a number of cases, for the growth of crystals from the gas phase and for the crystallization of organic substances, formulas (1) and (2) satisfactorily describe the experimental relationship between growth rate and supercooling. Here good quantitative agreement is usually found in the region of relatively high supercoolings [3, 5]. For small ΔT the growth rate calculated from this formula is considerably smaller than that observed [3, 6-8], since the actual crystal contains various kinds of defects making growth easy. Burton, Cabrera, and Frank [8] showed that the presence of screw dislocations in the crystals eliminates the necessity of forming two-dimensional nuclei, so that growth can take place even for extremely small supercoolings. Having used dislocation theory for the case of growth from the melt, Hillig and Turnbull [9] obtained the following expression for the growth rate:

$$V = A (\Delta T)^2. \tag{3}$$

It is shown in [10] that the degree of ΔT must lie between 1.6 and 1.8 if the thermal interaction between neighboring turns of the spiral is taken into consideration.

In certain cases this relationship agrees satisfactorily with experimental data. In particular, the following result is deduced in [9, 10] from an analysis of the data of Pollatshek [11] on growth rate in fine tubes:

$$V = 4 \cdot 10^{-4} (\Delta T)^{1.7} (\Delta T \leq 6° \text{ C})$$

and

$$V = 1.56 \cdot 10^{-5} (\Delta T)^{2.3} (\Delta T \geq 6° \text{ C}).$$

Several papers [5, 6, 11-15] have been devoted to a study of the temperature-dependence of the growth rate of salol, and the results of these differ.

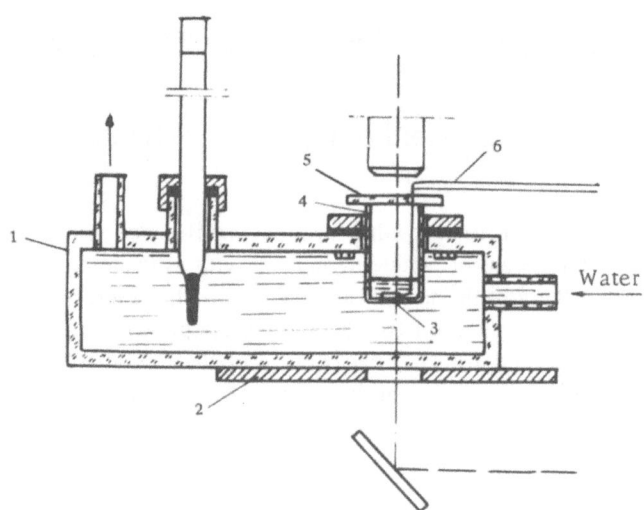

Fig. 1. Scheme of apparatus for studying the crystal growth of salol.

The paper of Danilov and Malkin [13] is deserving of special attention, as it describes a study specially undertaken to check the growth mechanism. The growth rate of the faces of individual crystals was examined on supercooling by 0 to 2.5°C. It follows from the data of [13] that in the supercooling range 0 to 1.5°C the crystals hardly grow at all, but at lower temperatures the rate becomes considerable and in a range of 1°C rises from $2 \cdot 10^{-5}$ to $5 \cdot 10^{-3}$ mm/min. The temperature-dependence of the growth rate is well described by Eq. (1). On the basis of this, the authors concluded that salol crystals grew by the formation of two-dimensional nuclei. Nevertheless, in another paper [6] devoted to experiments on individual salol crystals, a considerable growth rate was found for supercoolings of even 1° and less.

Iäntsch [5] considered that, for small supercoolings, crystals grow as a result of growth sites on the crystal faces, and for large supercoolings predominantly by the formation of two-dimensional nuclei. The theoretical curve for the temperature-dependence of the growth rate in salol, constructed on the assumption that the relative density of growth sites equals 0.38, agrees satisfactorily with experimental data [6, 11, 15]. The author affirms that for $\Delta T = 2.5$°C the growth is caused mainly by growth sites, while for larger supercoolings a considerable contribution comes from the formation of two-dimensional nuclei, and this later becomes predominant. Qualitatively, this conclusion would appear to be justified. As regards quantitative agreement between the entire theoretical curve and experimental data, this can hardly be considered satisfactory, since in the range of small supercoolings (up to 5°C) data from one group of authors [6, 11] are used, and in the range of large supercoolings those of others are used [15]; the experimental conditions of these may well have differed.

It follows from what has been said as well as from additional analyses to be found in the literature that for the same supercoolings, according to some papers, salol crystals have been found to grow both as perfect crystals and as imperfect ones. It is hard to determine whether this is a property of the crystals themselves or whether it is caused by other circumstances, such as different arrangements of the experiments, their conditions, and the care with which they were carried out.

In order to elucidate this, we attempted once more to study the temperature-dependence of the growth rate in the region of slight supercooling, where this is determined mainly by the growth mechanism, and the mobility of the molecules is less significant. In carrying out this investigation we measured not only the temperature of the bath but also that of the front; this was an extremely important factor and had not been done in the earlier investigations.

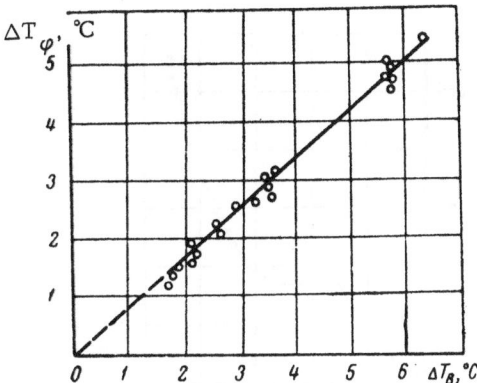

Fig. 2. Variation of the supercooling at the crystallization front (ΔT_F) with the supercooling calculated from the bath temperature (ΔT_B).

Method of Investigation

We studied the crystal growth of salol in the apparatus which is partially shown in Fig. 1. A transparent cuvette 1 is set on the object stand 2 of an MP-3 polarization microscope. The salol is placed in the thin-walled, flat-bottomed test tube 4. Water passes through the cuvette from an ultrathermostat. The temperature at the crystallization front was measured by a fine thermocouple 6 (hot-junction diameter 0.08 mm) introduced through the cover glass 5. The cold junction of the thermocouple was thermostated in a brass beaker by means of another ultrathermostat. The steady temperature in both the cuvette and the brass beaker was kept constant during the experiment to ± 0.01°C.

The emf of the thermocouple, proportional to the temperature difference between the hot and cold junctions, was amplified by a DC amplifier and recorded by an EPP-09 electronic potentiometer. For rapid changes of emf the amplified signal was recorded on a loop oscillograph. The thermocouple was calibrated with a Beckmann thermometer. The error of temperature measurement in the potentiometer record was ± 0.03°C. In all the experiments the linear crystallization rate (l.c.r.) was measured in individual salol crystals in transmitted polarized light. The salol in the test tube was melted and superheated by 5 to 7°C above the melting point, and was then supercooled to the temperature assigned and held at this temperature for an hour. Since spontaneous generation of crystals did not occur at the supercoolings in question, it was necessary to introduce a salol crystal seed 3 into the supercooled melt, placing it on the bottom of the test tube by means of a fine glass capillary. The l.c.r. was only measured for those crystals which were regularly faced and so oriented that the field of view of the microscope showed a regular rhomb with sides formed by the intersection of (100) and (110) planes. The velocity of motion of the opposite sides of the rhomb was determined by means of an ocular micrometer. The precision of measurement for the displacement of the crystallization front was ± 0.002 mm. When the crystallization front (side of the rhomb) touched the thermocouple junction, the temperature of the bath was noted, as was the temperature reading on the potentiometer.

The temperature of the liquid and crystal in our experiments was lower than at the phase-separation boundary. The temperature gradient in the liquid and solid phases fell with decreasing growth rate.

It was established in preliminary experiments that the thermocouple did not normally affect the growth rate nor bend the crystallization front, although in the individual cases when the front met the thermocouple there was an increase in the rate of growth. Only the rates measured before the crystallization front met the thermocouple were therefore used.

It is observed that the supercooling at the crystallization front (ΔT_F) depends on the supercoolings calculated from the bath temperature (ΔT_B) (Fig. 2). With decrease in bath temperature, the difference between the temperatures of the front and bath increases and reaches 0.8°C for $\Delta T_B = 5$°C. When $\Delta T_B = 1$°C, this difference is in fact 0.1°C, so that in the supercooling range 0 to 1°C the temperature of the front may be regarded as equal to that of the bath. Hence the l.c.r. relates to the bath temperature for supercoolings smaller than 1°C and to the temperature measured by the thermocouple for larger.

Determination of the Melting Point of Salol

In studying the mechanism of growth it is important to determine whether there exists a threshold supercooling below which, in accordance with the theory of the formation of two-dimensional nuclei, the growth of a perfect crystal is negligibly small. Since this threshold is relatively minute, it is necessary to determine

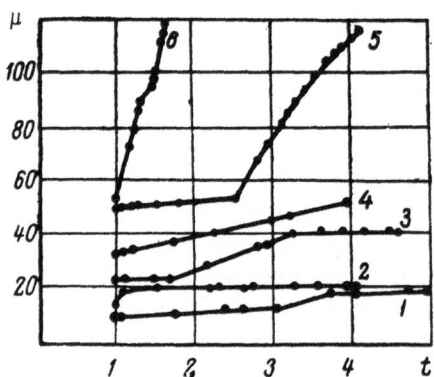

Fig. 3. Variation of the displacement of
the faces with time during growth of the
crystal: 1) 41.20°C; 2) 41.19; 3) 41.12; 4)
41.18; 5) 41.08; 6) 41.06°C.

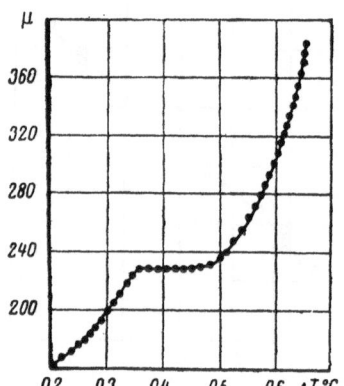

Fig. 4. Variation of crystal
growth rate with supercooling.
Cooling at constant rate of
0.1°C/h.

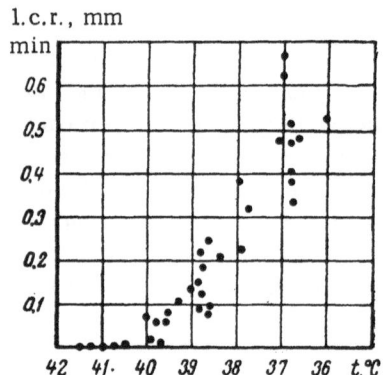

Fig. 5. Variation of crystal growth rate
with temperature.

the melting point accurately. Moreover, the smaller the temperature range of melting, the less impurities the given substance contains. The melting point of the salol under examination was determined on the same apparatus as was used to study the crystal
growth. The solid salol was held at constant temperature for
several hours. If a liquid phase was not observed under the microscope, then the temperature was raised by 0.1°C.

In our experiments, after holding at 41.35°C for 6 h, no
liquid phase was observed. At 41.41°C after the same holding
time a small quantity of liquid phase was observed. With the
raising of the bath temperature to 41.52°C, all the salol melted
in 40 min. When a crystallite was introduced into previously
melted salol at 41.46°C, it did not melt for 7 h. Hence a temperature of 41.5°C may be regarded as the melting point of salol.
This agrees closely with published data (41.6°C [7] and 41.4°C
[13]). The temperature range of melting did not exceed 0.17°C,
which indicated that the salol was satisfactorily free from
soluble impurities.

Discussion of Results

It was noted that the rate of growth of a particular crystal at constant temperature changed markedly with
time (Fig. 3). The starting positions of the curves in the figure are arbitrary and unconnected with the original
dimensions of the crystals, so that in comparing them one must take differences along the axis of ordinates, and
not absolute values. We see from Fig. 3 that one crystal (curve 5) grows at a rate of $2.8 \cdot 10^{-5}$ mm/min at
41.08°C for the first 1.5 h, and then the rate rises to $5.8 \cdot 10^{-4}$ mm/min. Two other crystals (curves 1 and 3)
behave analogously for 3 to 3.5 h, and then the rate again falls. Some crystals (curve 4) have a constant growth
rate during the whole time of observation. For large growth rates the nonuniformity becomes less noticeable,
and our measurements give an average rate. The nonuniform variation is also observed on slow cooling (Fig. 4).
A large set of growth rates for a given supercooling was also found in [7] for benzophenone crystals.

Variation of Crystal-Growth Rate with Supercooling

ΔT, °C	V, mm/min	ΔT, °C	V, mm/min	ΔT, °C	V, mm/min	ΔT, °C	V, mm/min	ΔT, °C	V, mm/min
0.10	no growth	0.30	no growth	0.35	$5.7 \cdot 10^{-4}$	0.44	$1.95 \cdot 10^{-3}$	0.69	$5.9 \cdot 10^{-4}$
0.17	»	0.30	$3.2 \cdot 10^{-4}$	0.38	$1.2 \cdot 10^{-5}$	0.52	$9.0 \cdot 10^{-4}$	0.69	$8 \cdot 10^{-3}$
0.17	$4.8 \cdot 10^{-5}$	0.30	no growth	0.38	$1.33 \cdot 10^{-4}$	0.55	$1.6 \cdot 10^{-3}$	0.80	$3.9 \cdot 10^{-3}$
0.18	$2.7 \cdot 10^{-4}$	0.30	$1.84 \cdot 10^{-4}$	0.38	no growth	0.55	$6.4 \cdot 10^{-3}$	0.80	$4.5 \cdot 10^{-3}$
0.18	$3.1 \cdot 10^{-5}$	0.30	$3.2 \cdot 10^{-4}$	0.38	$1.5 \cdot 10^{-5}$	0.56	$7.5 \cdot 10^{-4}$	0.87	$1.19 \cdot 10^{-2}$
0.25	no growth	0.30	$4.3 \cdot 10^{-5}$	0.40	$2 \cdot 10^{-3}$	0.58	$1.55 \cdot 10^{-3}$	0.90	$3.58 \cdot 10^{-3}$
0.25	$2.2 \cdot 10^{-4}$	0.30	$5.2 \cdot 10^{-5}$	0.40	$2.3 \cdot 10^{-3}$	0.60	$2.5 \cdot 10^{-3}$	1.00	$8 \cdot 10^{-3}$
0.26	$0.9 \cdot 10^{-5}$	0.30	$3.3 \cdot 10^{-5}$	0.42	$1.3 \cdot 10^{-3}$	0.60	$3.6 \cdot 10^{-3}$	1.00	$8 \cdot 10^{-3}$
0.26	$1.5 \cdot 10^{-4}$	0.32	$8.9 \cdot 10^{-5}$	0.42	$1.7 \cdot 10^{-3}$	0.68	$9.92 \cdot 10^{-3}$	1.03	$1 \cdot 10^{-2}$
0.27	$3.4 \cdot 10^{-4}$	0.35	$14 \cdot 10^{-5}$	0.42	$2.86 \cdot 10^{-5}$	0.68	$7.3 \cdot 10^{-3}$	1.38	$4.6 \cdot 10^{-2}$
0.28	$8.3 \cdot 10^{-5}$	0.35	$7.2 \cdot 10^{-4}$	0.42	$5.8 \cdot 10^{-5}$	0.68	$1.03 \cdot 10^{-2}$	1.38	$3.2 \cdot 10^{-2}$
0.30	$2.7 \cdot 10^{-4}$,	0.35	$7.3 \cdot 10^{-5}$	0.44	$6.3 \cdot 10^{-4}$	0.68	$2.9 \cdot 10^{-3}$		

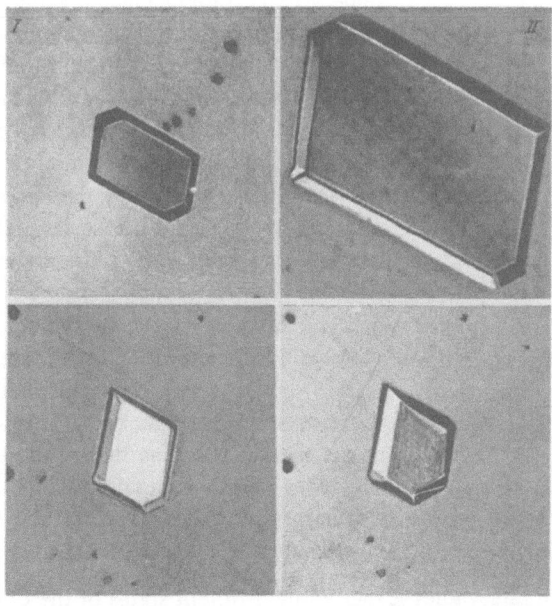

Fig. 6. Changes in crystal size over 3 h under the same conditions.

Fig. 7. Spirals on a salol crystal.

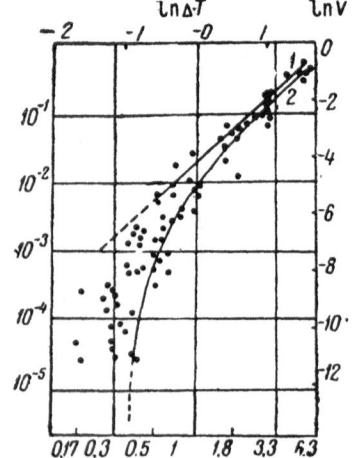

Fig. 8. Variation of ln V with ln ΔT.

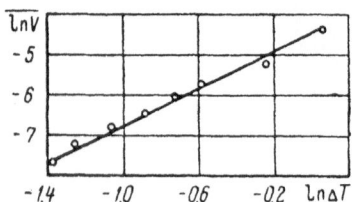

Fig. 9. Variation of ln V with ln ΔT for
the same crystal.

In studying the temperature-dependence of the l.c.r., measurements were made on crystals with both equilibrium and non-equilibrium rates of growth. The results obtained appear in Fig. 5, and those for small supercoolings are shown in the table. As we see, for the same supercoolings there is a large selection of growth rates, increasing as the supercooling diminishes. Thus, for example, for a 0.42°C supercooling the rate for different crystals varies from $1.7 \cdot 10^{-3}$ to $2.8 \cdot 10^{-5}$, i.e., by almost 60 times, and for $\Delta T = 0.36$°C, as well as crystals growing at a considerable rate ($1.3 \cdot 10^{-4}$ mm/min), there is one which remained constant in size for 10 h. For $\Delta T > 0.38$°C we found no crystals which failed to grow at all, but for smaller supercoolings such cases arose frequently (see the table). In the majority of cases, however, crystals grew at a considerable rate, even for $\Delta T = 0.18$°C. The variation in growth rate is shown in Fig. 6, from which we see that crystal 2 hardly altered its size, while crystal 1 increased by a factor of two.

The fact that for the same supercoolings some crystals failed to grow while others grew at varying rates provides a basis for the supposition that the salol crystals in our experiments had different degrees of perfection. The crystals which failed to grow at slight supercoolings were evidently the more perfect, while those growing at considerable rates under the same conditions contained faults, presumably of a dislocation character. This is confirmed by direct observations of the surface of individual crystals, among which some had smooth faces and others spiral (Fig. 7). We may also suppose that in the process of growth the structure of the face undergoes certain changes leading to a change in growth rate. One possible cause of this may be the capture of impurity atoms or insoluble particles capable of producing dislocations and hence an increase in the growth rate of the growing faces (Fig. 3, curves 1, 3, 5). An increase in growth rate may also be produced by a reduction in the work of formation of two-dimensional nuclei in the active sites formed by the re-entrant angles between a crystal face and the walls of the vessel and also the surfaces of insoluble impurity particles. When the crystal grows out to such an active site, the rate of formation of two-dimensional nuclei increases, and hence the crystal grows faster. After the crystal has grown past the active site, the growth rate resumes its original value (Fig. 3, curves 1 and 3). Confirmation of this clearly comes from the fact that, as a rule, the growth rate of crystals floating freely in the melt is less than that of those growing at the bottom of the test tube. If a floating crystal is slightly deformed by pressing from above with a glass capillary, its growth rate sharply increases. The fall in growth rate with time is apparently caused by the "poisoning" of the growth sites by adsorbed atoms [10] dissolved in the salol.

Thus the scatter in growth rate values is caused by different states of the crystal surfaces, and hence the growth of difference crystals may take place by means of different mechanisms. Hence, in order to explain the temperature-dependence of the growth rate physically, it would be incorrect to average all the observed rates for given supercoolings. Since the growth of imperfect crystals is more probable at slight supercoolings, it is natural to suppose that the maximum growth rate corresponds to the dislocation mechanism. The correctness of this assumption is confirmed by the straight-line relationship between the logarithm of the maximum growth rates and the logarithm of the supercoolings (Fig. 8, line 1). The relationship is described by the equation

$$V = 4.6 \cdot 10^{-5} (\Delta T)^2,$$

which agrees closely with (3). A similar relation is also found for an individual crystal (Fig. 9), the growth rate of which was measured during slow cooling.

The minimum cooling rates, as seen from Fig. 8, deviate considerably from a straight line (curve 2).

Analysis shows that the temperature-dependence of the minimum growth rates is better described by Eq. (1) than by (3). We may therefore suppose that some of the rates observed correspond to growth caused by the mechanism of two-dimensional nucleus formation. The values of $K = K_1 e^{-K_2/T}$ and K_3 found from the relationships between ln and $1/T(\Delta T)$ respectively, equals 0.3 and 1100. The specific peripheral energy calculated from K_3 ($\rho = 2 \cdot 10^{-7}$ erg/cm) agrees with the data of [13, 16]. The value of K, however, is much smaller than in [13].

Our estimate of the constants must be regarded as very approximate, owing to the comparatively small amount of data. Moreover, it would appear that not all the values of the extremal growth rates correspond to the proposed mechanism. It could be that some of the lowest rates are due to a dislocation mechanism, the action of which is retarded owing to adsorbed impurities. Finally, we must acknowledge that for large supercoolings (3 to 5°C), when the difference in growth rates is relatively slight (difference of a factor of 1.5 to 2) and the action of both mechanisms may be involved, the division into maximum and minimum growth rates is rather indefinite, and this introduces difficulties in constructing the temperature-growth rate relations and hence also in evaluating the corresponding constants.

Comparing the results obtained with the data of [13], we must point out that, for the same range of supercoolings, our growth rates are considerably higher. Moreover, we were never able to observe crystals failing to show growth at $\Delta T > 0.40$°C, whereas in [13] this threshold was 1.5°C. The reasons for this discrepancy are still not clear. It may be that it is associated with the presence of different impurities in our salol and that used in [13], these affecting the peripheral energy and hence the work of nucleus-formation to different extents, or else with the fact that the authors of [13] were able to obtain perfect crystals by some means which they failed to specify.

Despite the quantitative discrepancy, our data on the existence of a supercooling threshold and on the temperature-dependence of the growth rate constitute qualitative confirmation of the conclusions in [13] regarding the possibility of growing salol crystals by forming two-dimensional nuclei.

A comparison of the growth rates at $\Delta T < 1.5$°C with the theoretical curve of [5] shows that the maximum rates obtained are five to seven times smaller than those calculated by Iäntsch. This theory can thus apparently not be used, with the assumptions made in it, to describe the results obtained.

Conclusions

1. We have studied the temperature-dependence of the growth rate of salol crystals in the supercooling range 0 to 6°C.

2. We have shown that, for the same degrees of supercooling, different crystals grow at different rates. For supercoolings up to 0.4°C some crystals hardly grow at all, while others under the same conditions grow at considerable rates ($3 \cdot 10^{-4}$ to $2 \cdot 10^{-3}$ mm/min). As the degree of supercooling rises, so does the growth rate, and the difference between the growth rates of different crystals diminishes.

3. We have found that, for strictly identical conditions, the growth rate of a particular crystal changes with time. We have presented hypothetical reasons for this.

4. From an analysis of the temperature-dependence of the growth rate, we conclude that, under the super-cooling conditions studied, some crystals grow by a dislocation mechanism, and others by the formation of two-dimensional nuclei.

Literature Cited

1.   J. Gibbs, Thermodynamic Works [Russian translation](Gostekhizdat, Moscow—Leningrad, 1950).
2.   M. Volmer and Weber, Z. Phys. Chem. 119:227 (1926).
3.   H. Brandes, Z. Phys. Chem. 126:196 (1927).
4.   I. N. Stranskii and R. Kaishev, Usp. Fiz. Nauk 21(4):(1939).
5.   O. Iäntsch, Z. Krist. 108:185 (1956).
6.   H. Möller, Ctrbl. Mineralog. A 131 (1921).
7.   I. B. Morris and R. F. Strickland—Constable, Trans. Faraday Soc. 50:1378 (1954).
8.   W. Burton, N. Cabrera, and F. Frank, Elementary Processes of Crystal Growth [Russian translation] (IL, 1959), p. 5.
9.   W. Hillig and D. Turnbull, Elementary Processes of Crystal Growth [Russian translation] (IL, 1959), p. 293.
10.   A. A. Chernov, Usp. Fiz. Nauk 123:306 (1961).
11.   H. Pollatschek, Z. Phys. Chem. 142:289 (1929).
12.   A. L. Danilova and V. I. Danilov, Problems of Physical Metallurgy and the Physics of Metals. Collection (Institute of the Physics of Metals and Physical Metallurgy, Central Scientific Research Institute of Ferrous Metallurgy, 1949), p. 1.
13.   V. I. Danilov and V. I. Malkin, Zh. Fiz. Khim. 28:1837 (1954).
14.   G. Micus and U. Troltenier, Z. Phys. Chem. 2:229 (1954).
15.   K. Neuman and G. Micus, Z. Phys. Chem. 2(25):(1954).
16.   L. I. Chesnokov, Zh. Fiz. Khim. XXXVI:3 (1962).

CHARACTER OF THE LINEAR CRYSTALLIZATION
RATE/TEMPERATURE CURVE FOR HEXOACETATE

M. M. Mazhul' and L. K. Sharik

A number of organic substances with transparent melts, small linear crystallization rates, low melting points, and good capacity for supercooling, can conveniently be used in "model" experiments, since the laws of their crystallization under the influence of various physical factors may be carried over to the case of metals and other materials.

We set ourselves the problem of studying the crystallization properties of hexoacetate in order to use this for model experiments. We had to establish the melting point, obtain curves relating the linear crystallization rate (l.c.r.) v and rate of forming crystalline nuclei (r.f.c.n.) I to temperature, and determine the crystallization parameters of the substance.

For this purpose we selected the Tamman method, which was used by the authors of [1, 2, 5], with due allowance for the various failings in each specific case.

In order to prepare the samples we used capillary tubes with aperture diameter from 0.2 to 1.5 mm, and also plane parallel samples made by means of cover glasses.

In taking the experimental l.c.r./temperature curves, the necessary supercoolings were maintained with an ultrathermostat. The l.c.r. measurements were made on the microscope table by means of the specimen drive.

The melting point of the hexoacetate was determined by various methods and was found to be 98°C.

It was shown earlier by Tamman [1] that there is a temperature for which the l.c.r. is maximum, and this may be regarded as one of the characteristics of the substance. Later it was shown by Danilov [2] and others that the temperature for which the l.c.r. is maximum depends on the degree of purity of the substance.

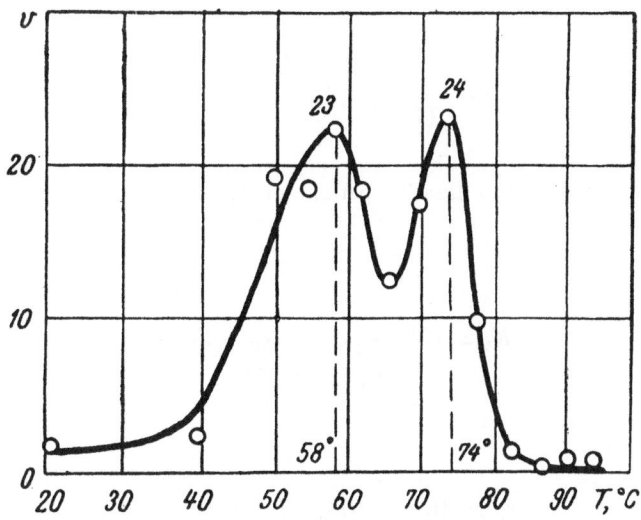

Temperature-dependence of the l.c.r. of hexoacetate.

Danilov noted that for substances in which the maximum l.c.r. was smaller than 1 mm/min the l.c.r./temperature curves featured a sharp maximum, while for substances with l.c.r. greater than 1 mm/min the l.c.r./temperature curves showed a region in which the l.c.r. did not depend on temperature (a "plateau").

It was shown by Meleshko [4] that the "plateau" and regions of instability of the l.c.r./temperature curve noted by Tamman and Danilenko did not in fact exist; this the author explained as the result of eliminating the effect of heat of fusion and certain other factors on the l.c.r.

The variation of the l.c.r. with the thickness of the sample in the capillary tubes or between the cover glasses was given in papers by Kondaguri, Gorskii [5], and Meleshko [4], and other authors, but the indication that the l.c.r./temperature curve could have more than one maximum appeared in Meleshko's paper [6].

This phenomenon is due to the genesis and development of nuclei belonging to several different modifications in the supercooled liquid. Each modification is characterized by its "own" functional relationships $v = v(\Delta T)$ and $I = I(\Delta T)$.

In our own study of the crystallization properties of hexoacetate, we found that the l.c.r./temperature curve had two maxima (see the figure). The distance between the maximum along the temperature axis was of the order of 8 to 10°C.

As seen from the l.c.r./temperature curves, the maximum l.c.r. for hexoacetate is approximately 0.2 mm/min. For this l.c.r. we may suppose that the temperature at the crystal−melt boundary is close to that of the ultrathermostat [2].

The fact that the l.c.r./temperature curve for hexoacetate has two maxima may be explained either, in accordance with [6], by the presence of nuclei of two modifications, or else, as in [5], by the role of surface layers of glass.

The character of the experimental l.c.r./temperature curve may be described, as we know, by Fulmer's formula

$$v = K_1 e^{-\frac{K_2}{T}} e^{-\frac{K_3}{T \Delta T}} \tag{1}$$

or the formula of Kaishev and Stranskii:

$$v = \frac{K_1}{(\Delta T)^2} e^{-\frac{K_2}{T}} e^{-\frac{B}{T \Delta T}}. \tag{2}$$

From Fulmer's formula we found the crystallization parameters K_1, K_2, and K_3 for hexoacetate, the mean values of these being:

For the first maximum

$$K_1 \approx 10^{15} - 10^{20}, \quad K_2 \approx 25 \cdot 10^3,$$
$$K_3 \approx 90 \cdot 10^3.$$

For the second maximum

$$K_1 \approx 10^{15} - 10^{20}, \quad K_2 \approx 50 \cdot 10^3,$$
$$K_3 \approx 70 \cdot 10^3.$$

The parameters were found by solving three equations with three unknowns.

Literature Cited

1. G. G. Tamman, Crystallization and Fusion (1903).
2. V. I. Danilov, Structure and Crystallization of Liquids(Izd. Akad. Nauk Ukr. SSR, 1956).
3. V. I. Danilov and A. I. Danilova, Probl. Fiz. Metall. i Fiz. Met. (1):80 (1949).
4. L. O. Meleshko, Study of the Crystallization Rate of a Single Grain as a Means of Measuring Perimetric Energy, Dissertation (1955); Zh. Fiz. Khim. 34(1) (1960).
5. F. K. Gorskii and G. L. Mikhnevich, Zh. Eksp. i Teor. Fiz. II(4):264 (1932).
6. L. O. Meleshko, Inzh.-Fiz. Zhurn. III(3):(1960).

STUDY OF THE TEMPERATURE-DEPENDENCE OF THE LINEAR
CRYSTALLIZATION RATE OF SALOL, BETOL,
SALIPYRINE, ANTIPYRINE, AND CODEINE

L. O. Meleshko

The first systematic investigations of linear crystallization rate (l.c.r.) were carried out by Gernez [1], who developed a method of studying l.c.r. based on measuring the velocity of displacement of the visible boundary between solid and liquid phases in glass tubes placed in a bath at appropriate temperature. Measuring the l.c.r. for sulfur and phosphorus at various temperatures of the melts, Gernez showed that the l.c.r. rose with increasing supercooling. The same conclusion was reached by Moore [2] in studying the crystallization of phenol and acetic acid. Further systematic studies of the crystallization of transparent organic melts carried out by Tamman and colleagues [3] showed that at the phase-equilibrium temperature the l.c.r. was zero. As the supercooling of the melt rose, the rate increased, and, after reaching a certain maximum value v_M at temperature T_M, remained constant over a wide temperature range ("plateau"), only falling again at extremely large supercoolings. Tamman divided substances into two groups, depending on the maximum growth rate. The first group contained substances with $v_M > 4$ mm/min. These are characterized by curves $v = f(\Delta T)$ with a "plateau" (Fig. 1a). The second group is composed of substances with $v_M < 4$ mm/min (Fig. 1b). For these the $v = f(\Delta T)$ curves have sharp maxima.

An analogous division was later (1949) established by Danilov [11]: curves with a "plateau" for $v_M > 3$ mm/min, and curves with a sharp maximum for $v_M < 3$ mm/min.

A repeatedly discussed question is how far experimental l.c.r. curves obtained by the method of Gernez correctly reflect the true $v = f(\Delta T)$ relationship, since ΔT is the difference in them between the melting point and the thermostat temperature, whereas as a result of the evolution of heat of crystallization the temperature at the phase boundary should be higher than in the thermostat [4-9]. To check this, a number of authors have measured the temperature at the phase boundary [4, 7-9]. It was found that, as the phase boundary intersects the thermocouple junction, there is a local rise in temperature, which does not reach the equilibrium values. The temperature rise depends both on the growth rate of the new phase and on the degree of supercooling of the melt: the larger these quantities, the more the junction is heated.

In the method of studying l.c.r. used by Gernez, Tamman, Danilov, and others, insufficient provision was made for heat-elimination, so that the heat of transformation had a considerable effect on the l.c.r. For these reasons, various authors at various times obtained contradictory results for the same substances. In view of this, papers devoted to l.c.r. have usually presented the $v = f(\Delta T)$ curves either schematically or, when curves with "plateaus" were in question, simply by presenting the figures corresponding to the maximum rate. Doubt has been repeatedly cast on this method of studying l.c.r. [22, 23].

Experimental $v = f(\Delta T)$ curves with sharp maxima (for glycerine) were compared with theory by Fulmer and Stranskii [10]. For curves with a plateau such a comparison was attempted by Danilov and Danilova [11]. Using the method of Pollatschek [8], Danilov and Danilova determined the temperature at the crystallization front of salol and plotted the curve relating the l.c.r. to the temperature at the crystal—melt boundary. However, the experimental curve referred to the temperature of the crystallization front retained its "plateau," and all the difficulties of its theoretical interpretation remained as before.

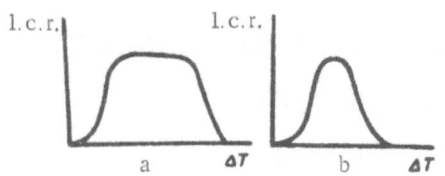

Fig. 1. Linear crystallization rate as a function of supercooling (Tamman's curves): a) for maximum l.c.r. greater than 4 mm/min; b) less than 4 mm/min.

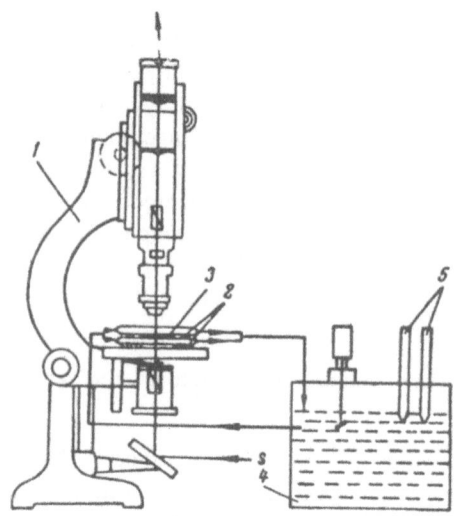

Fig. 2. Apparatus for studying linear crystallization rates of single grains in plane specimens: 1) polarization microscope; 2) plane—parallel glass cuvettes; 3) substance being studied; 4) ultrathermostat; 5)thermometers [12].

We have developed a new method of measuring the l.c.r. [12] which removes the main source of error in the Gernez method. In our approach, the substance under study is placed not in a capillary but between two glass plane—parallel cuvettes through which liquid circulates from an ultrathermostat (Fig. 2). The linear crystallization rate is measured from the growth of a single grain as the change in grain radius ΔR with time $\Delta \tau$. In the plane layer the crystal grows only to the side, and this reduces growth to a two-dimensional process, in which the possibility of convection flows does not exist. The liquid circulating through the cuvettes from the ultrathermostat maintains the layer at a constant temperature and ensures good heat-removal conditions. Temperature measurements at the crystallization front showed that in such flat specimens 0.3 to 0.35 mm thick the temperature at the crystal—melt boundary did not differ from the thermostat temperature by more than ± 0.1° over the entire temperature range of 2 to 90°C [14]. The transparent liquid did not render the observation of the growing crystal difficult, since on carefully purifying the liquid flowing through the cuvette it was impossible to decide visually whether it was moving. The walls of the glass cuvettes were about 0.2 mm thick, and their plane surfaces were carefully polished; before inserting the substance to be studied they were washed and dried in vacuum. The sample was fixed to the microscope table and kept in the same position during the entire period of examination (Fig. 2). In order to melt the sample, a liquid at a temperature some 10° above the melting point of the specimen was passed from an ultrathermostat through the cuvettes for 10 min. Then by means of a three-way stopcock the circulating hose was switched to the second ultrathermostat, and liquid at the temperature assigned to keep the required degree of supercooling in the melt was passed into the cuvettes. The temperature of the melt enclosed between the cuvettes was regulated by differential copper—constantan thermocouples in series with a mirror galvanometer. Control of the thermostat liquid was effected by mercury thermometers with 0.05° scale divisions. Our method of measuring l.c.r. from the growth of a single grain in plane specimens subjected to the flow of liquid from an ultrathermostat was used to determine the functional relationship $v = f(\Delta T)$ of five different organic substances: salol, betol, salipyrine, codeine, and antipyrine. The results are shown in Fig. 3. We see from the figure that, independent of the maximum growth rate, the $v = f(\Delta T)$ curves always have a sharp maximum; there is no "plateau" and no region of unstable values, such as were found by Tamman and Danilov. According to our method of investigation, the "plateau" is not only absent from salol, the v_M of which equals 4 mm/min, but even from such substances as antipyrine ($v_M = 39$ mm/min) and codeine ($v_M = 20$ mm/min), although in these v_M is considerably greater than the value of $v_M = 4$ mm/min for which the maximum in the $v = f(\Delta T)$ curves of Tamman, Danilov, and others always turned into a "plateau." Analogous results have recently been obtained by other workers using our method [13, 14].

By measuring the l.c.r. of various forms of nuclei in polymorphic betol crystals, it was established that each modification has its "own" $v = f(\Delta T)$ relationship. These differ from one another in the size and siting of the maxima, the equilibrium temperature, and the general form of the curves (Fig. 3d).

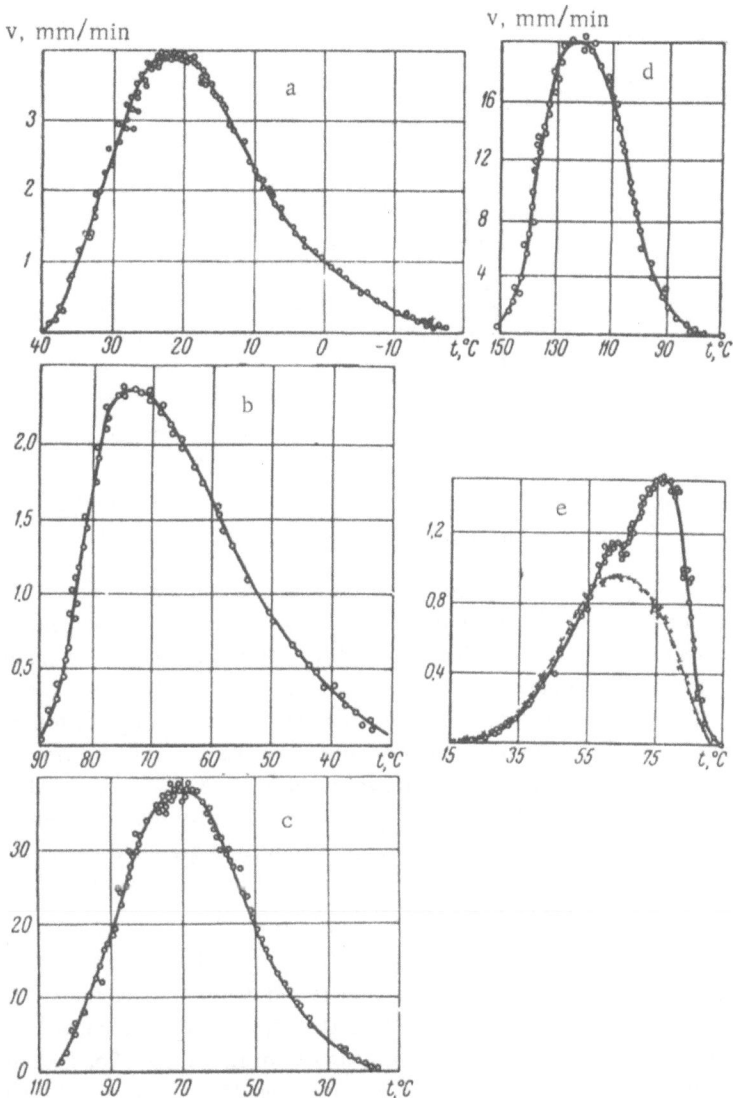

Fig. 3. Temperature-dependence of l.c.r. for a) salol; b) salipyrine; c) antipyrine; d) codeine; e) different forms of betol.

A few papers [10-12, 16] have been devoted to comparing the experimental l.c.r./temperature curves with the molecular—kinetic theory. In a number of cases this comparison was made over narrow temperature ranges in the rising or falling branch of the $v = f(\Delta T)$ curve [11, 16]. Comparison over the entire curve was made by Fulmer, Stranskii, and Kaishev [10], and by the authors [12]. The common failing in these comparisons was that the crystallization parameters U, D, and K of the formula $v = [K/(\Delta T)^n] \exp(-U/RT - D/T\Delta T)$ were regarded as constants, and in order to obtain the best agreement between theory and experiment an arbitrary choice of these constants was made. Actually, however, the parameters U, D, and K are functions of temperature. For organic compounds, the activation energy U rises sharply with falling temperature [17], while the parameter D characterizes the work of formation of a two-dimensional nucleus and hence also constitutes a function of supercooling. If the approximation relates to a small temperature range, then these quantities may very coarsely be regarded as constants. When we are considering the entire $v = f(\Delta T)$ temperature curve, however, this assumption is clearly inadmissible.

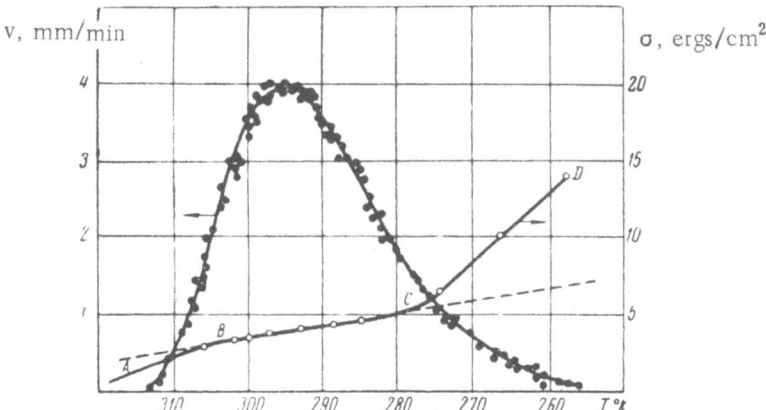

Fig. 4. Temperature-dependence of l.c.r. (v) and interphase
energy (σ) at the crystal—melt boundary for salol.

We proposed the following method, free from the disadvantages indicated, for comparing the experimental $v = f(\Delta T)$ curves with the formula $v = [K/(\Delta T)^n \eta \exp(-D/T\Delta T)$. From the experimental values of l.c.r. (v) and the viscosity of the supercooled melt (η), the relationship between $\ln(v\eta) + n \ln \Delta T$ and $1/T\Delta T$ is constructed. Then tangents are drawn to various points of the $\ln(v\eta) + n \ln \Delta T = f(1/T\Delta T)$ curve, the slopes of these giving values of D for different temperatures of the melt. This method of determining the temperature-dependence of parameter D, however, is not always possible, as it requires an experimental determination of the viscosity of the supercooled melt over the entire temperature range of the $v = f(\Delta T)$ curve, and this can only be done for substances with extremely weak crystallization capacities. This difficulty may be overcome by determining the temperature-dependence of the activation energy [18, 19], plotting graphically the relation between $\ln v + U/RT + n \ln \Delta T$ and $1/T\Delta T$, drawing tangents to this curve, and, as in the preceding case, using their slopes to determine the coefficients of $1/T\Delta T$ for various temperatures of the melt.

Analysis of the experimental $v = f(\Delta T)$ curve for salol (Fig. 3) with respect to the values of viscosity (η) and activation energy (U) for n = 0 gave identical values for D and different ones for K, in agreement with the computing scheme on which the formula was based. Since $D = \gamma S_B \xi^2 T_0/k\lambda$, the boundary tension ξ with respect to the perimeter of the two-dimensional nucleus can be found from its values. For a simple cubic lattice, there is an association between the specific perimetric energy at the boundary of the two-dimensional nucleus ξ and the specific surface energy σ:

$$\xi = \sigma (V/N)^{1/3},$$

where V is the molar volume and N is the number of molecules related to the gram-molecule of the crystalline substance. For more complex crystal lattices, this connection should be rather different, but the nature of the temperature-dependence of σ will probably not change. On this basis, we calculated the temperature-dependence of the interphase energy at the crystal—melt boundary for salol (Fig. 4).

As indicated by the foregoing, in comparing experimental data with theory we started from the assumption that the experimental $v = f(\Delta T)$ curves coincided exactly with the theoretical curves obtained from the $v = [K/(\Delta T)^n \eta] \exp(-D/T\Delta T)$. The principle was as follows. If reasonable values of ξ and σ were finally obtained and their temperature-dependence agreed with the general theoretical form, then this could be regarded as a qualitative confirmation of the theory. If not, then we could confidently affirm that there was no agreement between theory and experiment.

We see from the data presented (Fig. 4) that σ falls with increasing temperature, but in certain temperature ranges the slope is too large. The $\sigma = f(T)$ curve shown in Fig. 4 may arbitrarily be divided into three parts: AB, BC, and CD.

1. In region AB, close to the melting point, the value of σ is somewhat smaller than in the other two regions (BC and CD), which correspond respectively to the maximum and falling branch of the $v = f(\Delta T)$ curve, but the function $\sigma = f(T)$ rises sharply. This character of the $\sigma = f(T)$ relationship in region AB is apparently due to the fact that dislocations play a large part in crystal growth in this region [20, 21].

2. In region BC the $\sigma = f(T)$ relationship agrees satisfactorily with the general theoretical representation. Hence in this temperature range the crystal growth mechanism is probably close to that on which the molecular—kinetic theory is based.

3. For large supercoolings in the region CD, the temperature-dependence of the apparent quantity $\sigma = f(T)$ rises sharply.

In this region it would appear that there is a crystal formation mechanism differing from that described by the molecular—kinetic theory, so that the value of the surface energy σ cannot be calculated from the formulas of this theory. Clearly, the reason for this is the formation of the crystal by means of the addition of molecular groupings, rather than individual molecules alone.

Literature Cited

1. D. Gernez, Compt. Rend. 95:1278 (1882).
2. Moorse, Z. Phys. Chem. 12:545 (1893).
3. G. Tamman, Kristallisieren und Schmelzen (903).
4. I. I. Fridlyander, Z. G. Filippova, and M. S. Model', Izv. Sekt. Fiz.-Khim. Analiza Akad. Nauk SSSR (22):(1953).
5. H. Byrger, Proc. of the Section of Science (23):(1920).
6. W. J. Mason, The Physics of Clouds (Oxford, 1957).
7. S. C. Mossop, Proc. Phys. Soc. 68:(1955).
8. H. Pollatschek, Z. Phys. Chem. Abt. A 142(4):289 (1929).
9. H. A. Wilson, Phyl. Mag. 50:238 (1900).
10. I. N. Stranskii and R. Kaishev, Usp. Fiz. Nauk 21:4 (1939).
11. V. I. Danilov, Structure and Crystallization of Liquid (Izd. Akad. Nauk Ukr. SSSR, 1956).
12. L. O. Meleshko, Study of the Crystallization Rate of a Single Grain as a Means of Measuring Perimetric Energy (Minsk, 1954); Zh. Fiz. Khim. 34(1):(1960).
13. V. G. Zaremba, Pratsi Odes'k. Derzh. Univ., 148, 3, Prirodnie Nauki (1958).
14. G. L. Mikhnevich, Kinetics of Crystallization in Supercooled Organic Liquids and Supersaturated Solutions (Doctor's Dissertation) (Odessa, 1962).
15. L. O. Meleshko, Inzh.-Fiz. Zhurn. 3(3):(1960).
16. V. I. Malkin, Zh. Fiz. Khim. 28(11):(1954).
17. P. P. Kobeko, Amorphous Substances (Izd. Akad. Nauk SSSR, 1952).
18. L. O. Meleshko, Inzh.-Fiz. Zhurn. 3(9):(1960).
19. N. I. Shishkin, Zh. Tekhn. Fiz. 26(7):(1956).
20. G. W. Sears, J. Chem. Phys. (9):1630 (1955).
21. Collection: "Elementary Processes of Crystal Growth" [Russian translation] (IL, 1959).
22. B. V. Stark, I. L. Mirkin, and A. N. Romanskii, Collection of Proceedings of the Moscow Institute of Steel, No. 7 (1935).
23. Ya. S. Umanksii, N. S. Fastov, S. S. Gorelik, N. N. Finkel'shtein, M. E. Blanter, and S. T. Kishkin, Physical Metallurgy (Metallurgizdat, 1955).

METHOD OF DETERMINING THE TEMPERATURE-DEPENDENCE
OF THE NUMBER OF CRYSTALLIZATION CENTERS

L. O. Meleshko

For the determination of the temperature-dependence of the rate of crystallization-center development, Tamman [1] developed the following method. The melt to be studied was placed in a glass tube immersed in a thermostat to ensure the necessary supercooling and left there for a certain time. In the supercooled liquid there developed nuclei of the new phase, but these were not noticeable, as their growth rate in the temperature range used was small. The nuclei were made to appear by raising the temperature of the melt to a value close to the melting point. At this temperature no new crystallization centers appeared, but those formed earlier grew and became sufficiently visible to be counted. In this way Tamman determined the temperature-dependence of the rate of crystallization-center development (r.c.c.d.) for many organic substances. Analysis of the results showed that the function $I = \varphi(\Delta T)$ in all cases gave a curve with a sharp maximum. According to Tamman's data, insoluble impurities had a considerable influence on the formation of crystal nuclei, but the maximum on the temperature curve was not subject to the effect of the impurities.

Tamman's method for studying the number of crystallization centers (n.c.c.) proposed is not reliable. The principle of "revealing" the nuclei at a temperature close to the melting point distorts the true picture of the n.c.c./temperature curve. Since the dimensions of critical nuclei diminish with increasing supercooling, at a high "revelation" temperature nuclei which have not outgrown the critical dimensions corresponding to that temperature will melt. Only those which remain will grow to visible dimensions and be counted.

In order to study the effect of the revelation procedure on the rate of development of new-phase nuclei, we made a series of experiments, the most typical results of which appear below. The experiments were made by a previously described method [2]. After melting in boiling water, betol was held at the exposure temperature ($t_e = 0°C$) and then, after each holding time, "revealed" at temperatures of 15, 20, 25, 30, and 35°C. Each

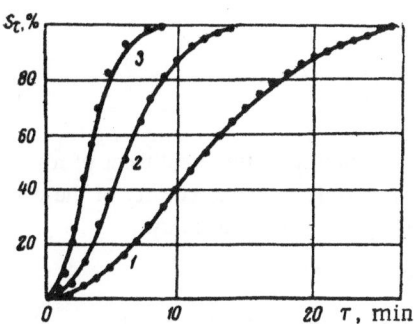

Fig. 1. Kinetic curves of betol (exposure 11 min at 0°C): 1) "revelation" temperature 25°C; 2) 30; 3) 35°C.

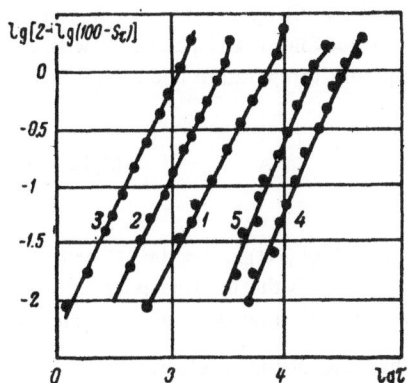

Fig. 2. Relationship between lg [2 − lg $(100 - S_T)$] and log τ for betol. (Exposure 0°C): 1) "revelation" temperature 15°C; 2) 20; 3) 25; 4) 30; 5) 35°C.

Time for the Transformation of Half the Mother Phase $\tau_{1/2}$, Constant b_1, and Density of Crystal—Nucleus Distribution I_0 for an Exposure Temperature of 0°C and Various Revelation Temperatures

t,° C	b_1	$\tau_{\frac{1}{2}}$	I_0
15	1,88	4860	93
20	2,06	2430	38
25	2,11	706	21
30	2,30	353	13
35	2,40	193	9

time, kinetic curves characterizing the total transformation rate under the given conditions were plotted at the revelation temperature t_p (Fig. 1). For the same exposure, as the revelation temperature rose, the mean density of the nuclei fell as a result of melting; this must affect the rate of transformation. The validity of these considerations may easily be checked from the kinetic curves $S_\tau = f(\tau)$ plotted for various revelation temperatures (Fig. 1).

From the data of the experimental investigations, we constructed the relationship between lg [2 − lg (100 − S_τ)] and lg τ (Fig. 2). We see from the figure that in the middle part of the transformation the experimental points fall neatly on straight lines, the slope of which gives the constant b for various revelation temperatures. The deviation of the experimental points from a straight line in the final transformation period is caused, first, by the imprecision of the formula $S_\tau = S_0 [1 - \exp(- aI_0 \tau_1^{b_1})]$, according to which the end of the process of crystallization must arrive at $\tau = \infty$, and, secondly, by random phenomena occurring during crystallization. Sometimes, in a particular part of the mother phase, the density of the crystal—nucleus distribution is considerably smaller than elsewhere. Hence the end of the transformation process may come considerably more slowly than theory demands.

The mean density I_0 of the initial transformation centers was calculated from the experimental data by the formula $I_0 = (\gamma_0 / \tau_{1/2})^{b_1}$. The table gives the results of calculating I_0 and also the half-transformation times used in the calculation.

Since the experimental conditions were such that the number of nuclei arising in the exposure period was constant for a definite series of experiments (same time and exposure temperature), the variation of I_0 can only be due to the melting of some of the crystals during revelation. On raising of the temperature, the melt passes into a thermodynamically new state with a correspondingly larger critical radius. Hence all the nuclei which at an exposure temperature had $(r_k)_e = 2\sigma M t_0 / \lambda \rho(t_0 - t_e)$ and were stable in the new conditions, could melt if

$$(r_k)_p = \frac{2\sigma M t_0}{\lambda \rho (t_0 - t_p)} > \frac{2\sigma M t_0}{\lambda \rho (t_0 - t_e)} = (r_k)_e,$$

where t_p, t_e, $(r_k)_p$, $(r_k)_e$ are the temperatures and critical radii corresponding to the conditions of revelation and exposure, t_0 is the melting point of the substance, M the molecular weight, ρ the density of the crystal, λ the heat of transformation, and σ the specific surface energy at the crystal—melt boundary.

Thus, use of the revelation procedure fails to give the function $I = \varphi(\Delta T)$.

We proposed a different method for determining the temperature-dependence of I and I_0 [2]. This reduces to an experimental study of the kinetics of isothermal crystallization at different temperatures and calculation of parameters I or I_0 from the formula $S_\tau = S_0 [1 - \exp(- \pi I_0 v^2 \tau_1^2 - (\pi/3) I v^2 \tau^3)]$ [3], depending on whether the revelation procedures were used.

The development of crystallization centers in a supercooled liquid is a purely chance phenomenon which is influenced by many factors not readily subject to exact calculation. Hence in order to obtain reliable data,

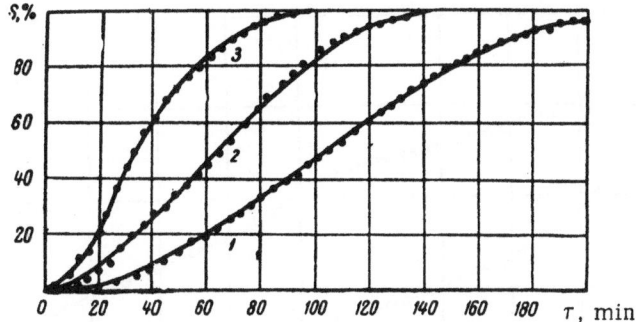

Fig. 3. Kinetic curves of betol obtained with the
revelation procedure: 1) crystallization temperature 0.8°C; 2) 25; 3) 30°C.

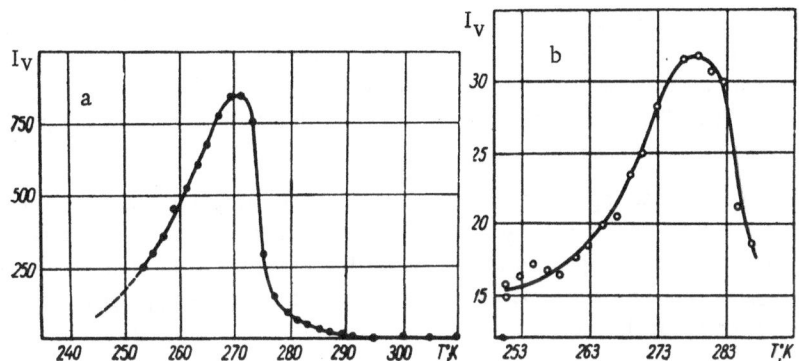

Fig. 4. Temperature-dependence of I obtained by taking kinetic curves
at various supercoolings of a) betol and b) antipyrine melts without the
revelation procedure.

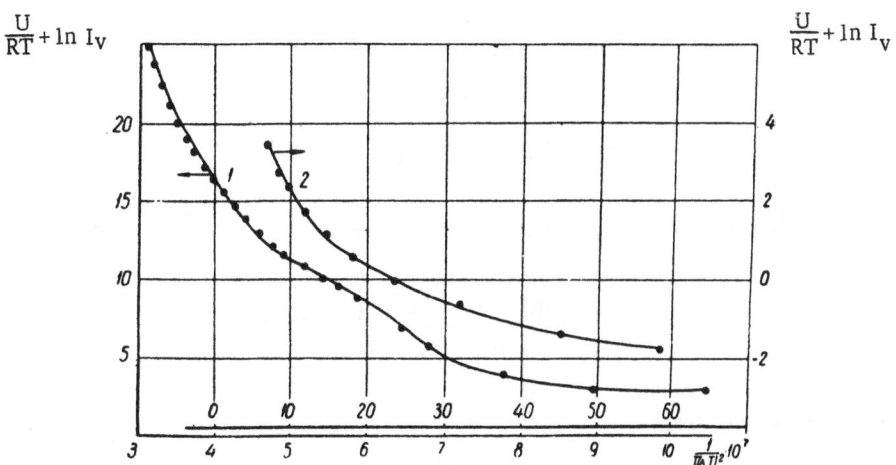

Fig. 5. Relation between $\ln I + (U/RT)$ and $1/T(\Delta T)^2$ for betol:
1) slight supercoolings; 2) large supercoolings.

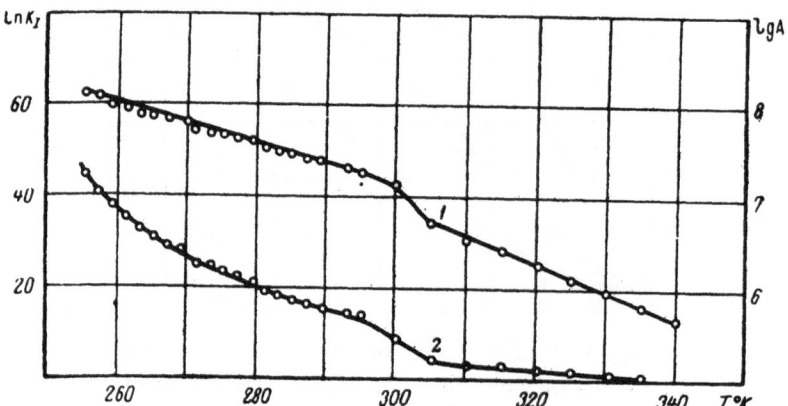

Fig. 6. Relation between lg $A = f(T)$ (1) and lg $K_I = f_1(T)$ (2) for betol.

the determination of the n.c.c. must be based on statistical methods in the presence of a large number of particles. In many experiments in which the Tamman method is used this condition was not satisfied. Experiments were usually repeated only a few times, the number of nuclei counted being comparatively small. In our own experiments this defect of the Tamman method was eliminated; the bulk formation and growth of a large number of nuclei were observed in a plane layer of fairly large area.

For the melting and supercooling of the substance, we used two Heppler ultrathermostats of the NB type: in one of these a temperature of approximately 10° above the melting point of the substance was maintained, and in the other the temperature corresponded to the supercooling assigned. In order to heat or cool the specimen, liquid of an assigned temperature was passed into the cuvette from the appropriate thermostat [2]. The parameters I and I_0 were calculated as follows.

From the time-law expression,

$$S_\tau = S_0 \left[1 - \exp\left(-a_0 I_0 \tau_I^{b_1} - a I \tau^b \right) \right] \tag{1}$$

[where for nuclei of circular form $a_0 = \pi v^2$ and $a = (\pi/3)v^2$] obtained on considering both spontaneous and forced crystallization [3], in the absence of the revelation procedure ($I_0 = 0$) we obtain

$$S_\tau = S_0 \left[1 - \exp\left(-a I \tau^b \right) \right]. \tag{2}$$

Hence

$$\alpha = S_\tau / S_0 = 1 - \exp\left(-a I \tau^b \right). \tag{3}$$

From (3) the transformation time of any part of the mother phase is

$$\tau = \left[\frac{2.201 \lg(1 - \alpha)}{v^2 I} \right]^{1/b} = \gamma I^{-\frac{1}{b}}. \tag{4}$$

Hence for half the transformed phase

$$I = \left[\frac{\gamma}{\tau_{50}} \right]^b. \tag{5}$$

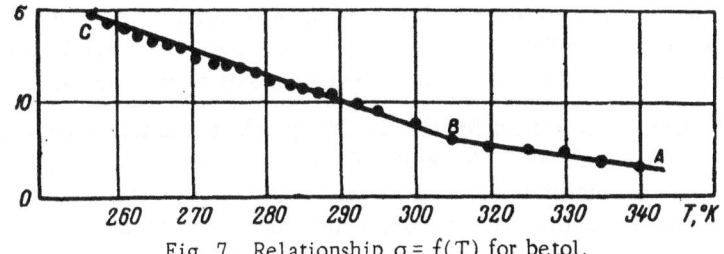

Fig. 7. Relationship $\sigma = f(T)$ for betol.

If the revelation procedure is used (which must be done with as large supercoolings as possible on the falling branch of the $v = f(\Delta T)$ cuve), then in formula (1) $I = 0$, and, in analogy with the above method, we have

$$I_0 = \left(\frac{\gamma_0}{\tau_{50}} \right)^b .$$

(6)

Thus from kinetic curves taken in isothermal conditions for various supercoolings of the melt we find I or I_0, depending on whether the revelation procedure is used. From the data obtained we construct the curves for $I = \varphi(\Delta T)$ or $I_0 = \varphi_1(\Delta T)$.

From (2) for betol at $\alpha = 1/2$ we obtained

$$I_{\frac{1}{2}} = 0.66/v^2 \, \tau^b_{\frac{1}{2}} .$$

(7)

The temperature-dependence of the rate of forming transformations centers in supercooled betol obtained from the kinetic curves (some of which appear in Fig. 3), without the revelation procedure and formula (7), is given in Fig. 4a. In the same way we found the $I = I(\Delta T)$ relationship for antipyrine (Fig. 4b).

Comparison with Theory

According to fluctuation theory, the probability for a new-phase nucleus to develop (I) in the supercooled liquid is given by

$$I = K_I \exp \left[-\frac{U}{RT} - \frac{A}{T(\Delta T)^2} \right] .$$

(8)

We compared the experimental curve $I = I(\Delta T)$ for betol (Fig. 4) obtained without revelation with formula (8), allowing for the temperature-dependence of the activation energy found earlier [4].

In order to determine the parameters A and K_I of formula (8), the relationship $\ln I + U/RT$ and $1/T(\Delta T)^2$ was constructed (Fig. 5).

As seen from the figure, this relationship forms a concave curve, the sharpness of which depends on the supercooling of the melt. Drawing tangents at various points of this curve, from their angles made with the axis of abscissas we find the value of A for different supercoolings of the melt. The results obtained are shown graphically in Fig. 6.

We see that the $\lg A = f(T)$ relationship is formed by two straightline sections with a bend in the region of 300 to 305°K. The line to the left of this point corresponds to the function $I = \varphi(\Delta T)$, where the probability of crystal nuclei developing is extremely great. In this region the nucleus-formation mechanism probably has a mainly homogeneous character. To the right of the bend, the main part in forming nuclei is apparently played by active impurities. As seen from Fig. 6, the $\lg A = f(\Delta T)$ relationship in the left-hand section, where use of formula (8) may be considered justified, constitutes a linear function of temperature.

If in an analogous way we analyze the experimental curve $I_0 = \varphi_1(\Delta T)$ for betol (using the revelation procedure), the relationship between $U/RT + \ln I_0$ and $1/T(\Delta T)^2$ is expressed by a straight line, from the slope of which we obtain $A = 119{,}871{,}790$. From the intercept made on the axis of ordinates we find $\lg K_I = 36{,}4246$ [4].

We determined the kinetic coefficient K_I from Eq. (8) by substituting the value of I, A, and U for corresponding temperatures of the betol melt. The results of the calculation appear graphically in Fig. 6. The values of K_I depend strongly on temperature. This appears most sharply in the left-hand side of the graph, where the nucleus-formation mechanism evidently has a homogeneous character.

The surface tension at the crystal—melt boundary may be found by using the value of A in formula (8).

For betol, we have from (8) [4]

$$\sigma = \sqrt[3]{\frac{3k\,\lambda^2\,A}{16\pi\,T_0^2\,v_{\text{B}}^2}}\,.$$

(9)

Inserting the values of A and v_{B} for betol in (9), we find the interphase energy σ for various temperatures of the melt. The results of the calculation are shown graphically in Fig. 7. We see from the figure that the function $\sigma = f(T)$ has a break. From the way in which $\sigma = f(T)$ varies, the whole range of supercooling studied may be divided into two regions.

Region AB. Here we suppose that there is a fluctuation mechanism for the formation of nuclei at active impurities. The monotonic character of the variation in $\sigma = f(T)$ in this section agrees with general theoretical considerations.

Region BC. Here the temperature-dependence of the expression is considerably sharper than in the region AB. One possible cause underlying this behavior of $\sigma = f(T)$ may be as follows.

The apparent value of $\sigma = f(T)$ may be associated with a law for the formation of new-phase nuclei differing from that described by the molecular—kinetic theory, which fails to allow for the structure of the liquid phase and the possibility of individual groupings of atoms taking part in the crystallization process.

Literature Cited

1. G. Tamman, Z. Phys. Chem. 25(3):441 (1898).
2. L. O. Meleshko, Inzh.-Fiz. Zhurn. 2(9):(1959); Collection: "Crystallization and Phase Transformations" (Minsk, 1962), p. 61.
3. O. M. Todes, Kinetic Processes of the Formation of a New Phase, Doctor's Dissertation (Moscow, 1944).
4. L. O. Meleshko, Inzh.-Fiz. Zhurn. 4(10):(1961).
5. Ya. I. Frenkel', Selected Works, Vol. 3, Kinetic Theory of Liquids (Izd. Akad. Nauk SSSR, 1959).

EFFECT OF CRUCIBLE MATERIAL AND THE PURITY OF THE ORIGINAL METAL ON THE SUPERCOOLING OF IRON

V. P. Kostyuchenko and D. E. Ovsienko

A most important characteristic of the crystallization capacity of pure metal is supercooling. This is also a criterion of the activity of introduced impurities. From the way in which the rate of forming crystallization centers depends on the supercooling, we may judge the mechanism underlying the influence of various impurities.

The supercooling of 150 to 500 g of iron at 260 to 270°C was achieved in [1, 2], and 10 to 100 μ diameter drops of iron were supercooled at 295°C in [3]; larger drops were supercooled in [4]. The various authors affirmed that the supercooling observed corresponded to the spontaneous crystallization of pure iron. It appears from [5], however, that, even for 40 to 50 μ diameter iron drops supercooled at 300 to 350°, crystallization centers arise on active parts of the substrate rather than spontaneously. Hence we may suppose that under certain conditions even larger supercoolings may be achieved by iron.

Moreover, in [1, 2] the maximum supercoolings of 260 to 270° were only obtained for relatively large volumes in individual cases, whereas in most of the experiments the supercooling was 100 to 150° or less. Thus, in order to study the effects of impurities, not only must we attain the maximum supercooling of the pure metal but we must also attain good reproducibility of the data, since otherwise the results may be indeterminate and their interpretation ambiguous.

It is known that supercooling is extremely sensitive both to impurities present in the metal itself and to products formed on melting as a result of action between the metal and the crucible walls or the atmosphere. The effects of these impurities frequently prove impossible to remove by ordinary deactivation processes, such as superheating or successive recrystallizations.

In order to attain the maximum stable supercoolings, we must create conditions which eliminate these factors as far as possible. In choosing these conditions, it is important first of all to know the effect of the crucible material, the degree of purity of the original material, the purity of the inert gas, and so on.

The aim of the present investigation was to elucidate the effect of these factors on the supercooling of iron. The experiments on the supercooling of pure iron were especially interesting because we had in our possession iron with a higher degree of purity than the iron used by earlier workers.

Results and Discussion

Effect of the Crucible Material on the Supercooling of Iron

The experiments were performed in a high-frequency system of the MVP-3M type. The crucible containing the metal was placed inside a quartz tube on a special ceramic stand (Fig. 1). In order to produce better thermal insulation and to protect the stand and the quartz from the action of the liquid metal if the crucible should break, the space between the walls of the crucible and the stand was filled with fine refractory particles. The sample-crystallization process was observed and the temperature measured through an observation window in the flange closing the upper part of the quartz tube.

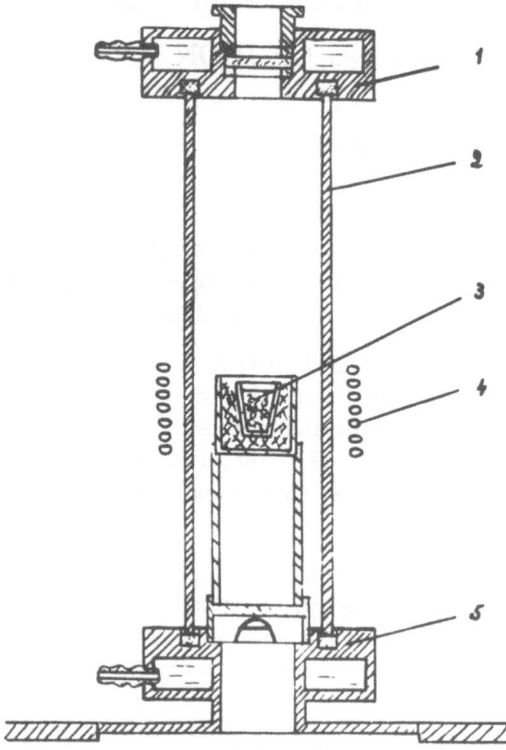

Fig. 1. System for determining supercooling: 1) upper flange; 2) quartz tube; 3) sample; 4) inductor; 5) lower flange.

The temperature was measured by an electronic color pyrometer with an ÉPP-09 recording potentiometer. The measuring accuracy was ± 10°. The apparatus, fresh from the factory, had a lower limit of 1340°, which would not accommodate iron supercoolings of more than 200°. By a suitable choice of light-filter glasses we were able to extend this limit to 1260°, thus widening the measurable supercooling range to 270 or 280° within the scale of the apparatus. Certainly, the measuring accuracy near the scale zero was somewhat reduced; in this case we only recorded the range 270 to 280°, indicated by a dash in the figures.

First, the metal in the solid state was degassed at 1000 to 1300°C in a vacuum of 10^{-3} to 10^{-4} mm Hg for 1 to 1.5 h, until gas evolution ceased. Previously purified argon was then introduced into the system, after which melting began. Cooling was effected at a constant rate of ~ 15°/sec.

For the investigation we chose crucibles made from magnesium oxide, beryllium oxide, zirconium dioxide, and aluminum oxide, i.e., from the materials most frequently used in melting iron and steel. The height of the crucibles was 20 to 25 mm and the diameter 19 to 22 mm. The weight of the ingot was 20 to 40 g. All the experiments with the various crucibles were carried out in identical conditions.

The subject of investigation in this series of experiments consisted of iron produced in the Central Scientific-Research Institute of Ferrous Metallurgy by the annealing and sintering of carbonyl iron powder in hydrogen. In purity, this was close to iron of the B_3 class and had a greater tendency to supercooling than any of the other types of iron at our disposal.

Table 1 and Figs. 2-4 present data for three samples out of each series of experiments (for each crucible material). In these figures the axis of abscissas gives the numbers of the samples and the number of successive fusion-crystallization measurements, and the axis of ordinates shows the superheatings (ΔT_+) upwards and the supercoolings (ΔT_-) downwards. As the data show, the supercooling depends substantially on the crucible material.

In magnesium oxide crucibles (see Table 1), supercoolings were usually not observed, except in individual measurements, where they were no more than 5 to 10°. Neither increasing the number of repeated recrystallizations nor heating to 500° above the melting point of iron (which in other cases led to deactivation of the impurities) changed the result. Ingots melted in these crucibles were usually pure and carried no traces of any con-

TABLE 1. Data on the Supercooling of Iron in Magnesium Oxide Crucibles

Sample	Superheating Supercooling	50 0	100 0	200 0	250 0	280 0	300 0	330 0
Sample	Superheating Supercooling	20 10	130 0	160 0	360 0	400 0	470 0	— —
Sample	Superheating Supercooling	150 0	200 0	300 0	330 0	350 0	500 5	500 0

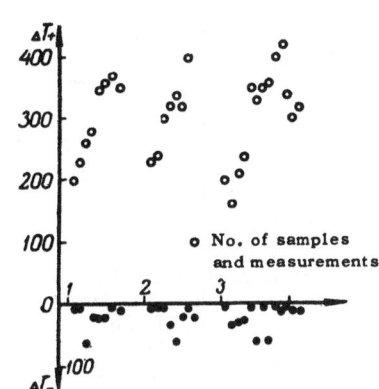

Fig. 2. Supercooling of iron in crucibles made from beryllium oxide.

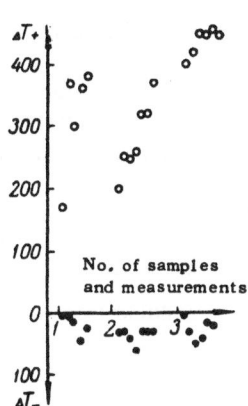

Fig. 3. Supercooling of iron
in crucibles made from zirconium dioxide.

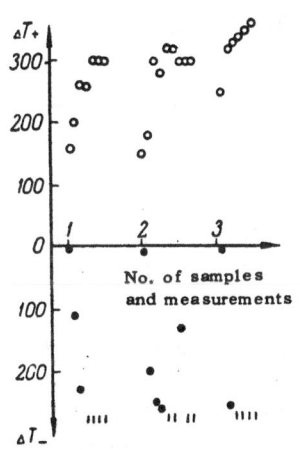

Fig. 4. Supercooling of iron
in crucibles made from
aluminum oxide.

tamination. The inner surface of the crucible in which the ingot lay, however, was covered by a dark, pinkish deposit, apparently formed by interaction
between the crucible and the iron or its oxides.

In experiments with beryllium oxide crucibles, as seen from Fig. 2,
supercoolings reaching 60° occurred. In the majority of cases, however, they
lay between 0 and 30°. This range did not alter for superheatings up to 450°
and 20-min holdings at these temperatures. The resultant ingots were also
pure, although the surface was rather lacking in luster. After melting, the
crucible was the same as it was before, and showed no trace of interaction
with the metal.

Almost the same results were obtained with crystallization in crucibles
of zirconium dioxide (Fig. 3). In contrast to the meltings in beryllium oxide
crucibles, however, there were clear traces of interaction between crucible
and ingot. The surface of the ingot had a yellow deposit and impregnation,
and the crucible was eaten away where the ingot had been.

Quite different results were obtained on crystallizing iron in aluminum
oxide crucibles (Fig. 4), where stable and reproducible supercoolings of 270
to 280° were easily reached. Such supercoolings were found on superheating
by 150 to 200°. Superheating by 500° and more with considerable holding times (up to 20 min) did not alter
the result. Ingots melted in these crucibles had a mat surface with clearly discernible grain structure where the
liquid surface had been. The side surface was covered in places with a dark deposit, indicating interaction between metal and crucible. The walls of the crucible were impregnated with iron or some of its compounds and
had become dark. Nevertheless, this obvious interaction between the metal or its compounds and the crucible
material not only did not prevent supercooling but evidently aided it.

Thus, of all the materials studied only the aluminum oxide crucibles were able to give iron supercoolings
of 270 to 280°. The earlier investigations [2, 4, 5] which achieved large supercoolings also used crucibles or
substrates of aluminum oxide. In [1] also, high degrees of supercooling were achieved in aluminum oxide crucibles, although the authors also obtained these supercoolings in other crucibles which they termed "oxide" without specifying their composition.

TABLE 2. Impurity Content in Types of Iron Examined, %

Type of iron	C	Si	Mn	S	P	Other impurity
Carbonyl A_1	$1—4 \cdot 10^{-2}$	$1 \cdot 10^{-2}$	$1 \cdot 10^{-2}$	—	—	
Carbonyl A_2	$1—2 \cdot 10^{-2}$	$1—2 \cdot 10^{-2}$	$1 \cdot 10^{-4}$	—	—	$Cu = 0.5—1 \cdot 10^{-4}$; $Ni = 8 \cdot 10$; No Pb
Electrolytic	$2 \cdot 10^{-2}$	$1.7 \cdot 10^{-2}$	$9 \cdot 10^{-4}$	$2.3 \cdot 10^{-2}$	$3.2 \cdot 10^{-3}$	
CSRIFM iron	$3 \cdot 10^{-2}$	Traces	Traces	$2 \cdot 10^{-2}$	Traces	$N_2 = 2 \cdot 10^{-3}$, Cr, Al, Mo, Ti, Pb, Co not observed

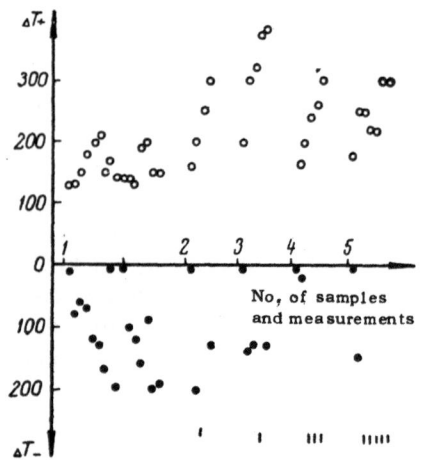

Fig. 5. Supercooling of various types of iron.

The data obtained are clearly insufficient to draw a definite conclusion on the reasons for such different effects of the crucible material on the supercooling of iron. Published data (which simply indicate that the above-considered crucible materials interact weakly with iron but strongly with iron oxides) are also insufficient for this. We can thus only make very general observations.

It may be that the crucible material dissolves to some extent in the iron and acts to different degrees as a surface-active impurity. It may also be that the refractory materials studied or the products of their reaction with the metal have different degrees of crystallochemical correspondence with the iron lattice and different activities in respect of the development of crystallization centers on them.

It is not impossible that in the case of aluminum oxide there is an interaction with oxides of iron, and that the product of this reaction is far less active than the iron or aluminum oxides themselves, these being well known to reduce the supercooling of iron [1, 2, 5]. There are a number of cases in the literature in which additives have neutralized impurities which formerly sponsored crystallization and have simultaneously sharply increased supercooling [6, etc.].

In order to check the ideas discussed and elucidate the actual mechanism of the effect of crucible material on the supercooling of iron, special experiments and detailed theoretical analyses are required.

Effect of the Degree of Purity of the Iron and the Inert Atmosphere on the Supercooling

Investigations were made on relatively pure types of iron, carbonyl class A_1, carbonyl class A_2, electrolytic (Tul'sk factory), and iron produced in the Powder Metallurgy Laboratory of the Central Scientific-Research Institute of Ferrous Metallurgy (CSRIFM) by deep refining of carbonyl powder in hydrogen. Data on the purity of the types of iron cited appear in Table 2.

The experiments were conducted in an argon atmosphere, the metal having been previously degassed in aluminum oxide crucibles. The method remained as before, except for the fact that the weight of the melt was increased to 100 g. The argon was purified in a special siphon by repeated circulation through metallic calcium heated to 700°C. The cooling rate was 10°/sec.

The results showed that different types of iron tended to supercool to varying degrees. Figure 5 shows comparative data for five samples of the types of iron examined. As seen from the figure, for carbonyl iron class A_1 (sample 1) the supercooling gradually increases as the number of meltings and crystallizations rises, and after nine or ten remeltings at superheatings of 150 to 200° it reaches 200°. Further superheating by 300 to 400° and repeated recrystallization does not increase the supercooling. In addition to the maximum super-coolings of 200°, smaller values are constantly observed.

After 90-min annealing in hydrogen at 1200°, the same iron began to supercool in a more satisfactory manner, reaching in individual cases as high as 260 to 280° (sample 2). The overwhelming majority of super-coolings, however, lay within the range 100 to 200°. Approximately the same result was obtained for carbonyl iron class A_2 (sample 3), though in this case the maximum supercoolings were reached rather more easily.

The best results were obtained for electrolytic iron and the iron produced in the CSIRFM (samples 4 and 5). Both types of iron behave alike: they easily supercool by 270 to 280° even after the first two or three remeltings for comparatively small superheatings (200 to 300°). These supercoolings reproduced themselves well both for single and different samples; the number of these exceeded fifteen.

These results, however, were obtained only when the melting took place in highly purified argon. In or-dinary factory argon, which contains (according to specification (0.003% oxygen, 0.05% nitrogen, and 0.01% carbon, supercoolings even for the purest iron samples did not exceed 100 to 150°.

Experiments on the supercooling of the types of iron discussed showed that in all cases the degree of purity of the argon had a greater effect than variations in the impurity composition of the original metal. The fact that small variations in the main impurity content of the iron have little effect on the supercooling is also in-dicated by the circumstance that the maximum supercoolings showed few differences for the last three types of iron (see Table 2). It is true, as we have already indicated, that the types of iron studied were of fairly high purity.

The reduction in the supercooling of the purer types of iron on melting in unpurified argon is evidently connected with the presence of oxygen, which forms iron oxides, since purification of the gas is aimed chiefly at oxygen. Apart from this, it is shown in [1] that nitrogen does not affect the supercooling of iron. The presence of iron oxides, however, as shown in [5], reduces the supercooling to 100 or 150° and less, just as on melting in an unpurified atmosphere. It may be that the relatively small and unstable supercooling of carbonyl iron A_1 is also connected with the presence of oxides, since after it is reduced in hydrogen the supercooling improves.

Clearly, of all the impurities in the types of iron studied, it is oxygen which has the greatest effect on supercooling.

Conclusions

1. We have shown that supercooling depends substantially on the crucible material and the degree of purity of both the metal and the inert atmosphere. The greatest and most reproducible supercoolings of 270 to 280° are attained for electrolytic iron and iron produced in the CSRIFM by refining in hydrogen, with melting in a well purified inert atmosphere in aluminum oxide crucibles.

2. We have suggested that the reduction in supercooling on melting in an insufficiently pure inert atmos-phere is associated with the formation of iron oxides.

Literature Cited

1. P. Bardenheuer and R. Bleckmann, Mitt. Kais. Wilg. Inst. Eisenforsch. 21(13):201 (1939).
2. D. S. Kamenetskaya, É. P. Rakhmanova, E. Z. Spektor, and V. I. Shiryaev, Probl. Metalloved. i Fiz. Met., No. 6 (Metallurgizdat, 1959).
3. D. Turnbull and R. Cech, J. Appl. Phys. 21(8):804 (1950).
4. A. I. Dukhin, Probl. Metalloved. i Fiz. Met. No. 6 (Metallurgizdat, 1959).
5. D. E. Ovsienko and V. P. Kostyuchenko, Vopr. Fiz. Met. i Metalloved. No. 10 (Izd. Akad. Nauk Ukr. SSSR, 1960); Growth of Crystals, Vol. III (Izd. Akad. Nauk SSSR, 1961); English translation: New York, Con-sultants Bureau, 1962.
6. V. E. Neimark, Probl. Metalloved. i Fiz. Met. No. 7 (Metallurgizdat, 1962).

BROADENING OF THE REGION OF PRIMARY SOLID SOLUTIONS
IN ALLOYS OF EUTECTIC AND PERITECTIC TYPES

I. S. Miroshnichenko

During the crystallization of alloys of the eutectic and peritectic types, primary solid solutions with limited solubility of one component in the other are formed.

As was shown earlier [1-4], on the crystallization of certain alloys (Ni−C, Co−C, Al−Mn, Al−Cr, Al−Ti, Al−V, Al−Fe, Sn−Sb) with cooling rates up to 10^6 deg/sec, there is a broadening of the region of solid solutions. The maximum solubility as given by the equilibrium phase diagram is substantially increased (see table).

At the same time one finds many binary systems (Pb−Cd, Al−Si, Sn−Bi Al−Ge, Al−Zn) in which there is no broadening of the region of solid solutions under the same crystallization conditions. The solubility lies within the limits of the equilibrium phase diagram.

All the investigated systems in which the primary solid-solution region undergoes broadening contain intermediate phases. Moreover, the development of strongly supersaturated solid solutions is accompanied by the crystallization of metastable intermediate phases.

The substantial rise in the solubility of the primary solid solutions as the cooling rate of the crystallizing melt increases takes place not gradually but suddenly, on reaching a certain critical cooling rate. Thus, for example, in Ni−C and Co−C alloys (Fig. 1) containing a sufficient amount of carbon ($> C_{M1}$), solid solutions of only two compositions crystallize: C_{M0} and C_{M1}. The first of these is formed for small cooling rates and the second for large. In some intermediate range of cooling rates, the simultaneous formation of two solid solutions with different solubilities (C_{M0} and C_{M1}) is possible. The x-ray diffraction photographs in this case clearly show two systems of lines belonging to the two solid solutions (Fig. 2). The alloys also contain stable graphite and metastable carbide.

Degree of Supersaturation of Primary α-Solid Solutions on Crystallization at a High Cooling Rate

Solubility	Systems							
	Ni−C	Co−C	Al−Mn	Al−Cr	Sn−Sb	Al−Ti [5]	Al−V	Al−Fe
Max. equilib., wt. %	0.6	0.8	1.4	0.7	10.5	0.28	0.37	0.05
At high cooling rates, wt. %	1.85	1.65	9.8	5.8	16.9	0.32	1.17	0.17
Increase on max. equilib. solubil., times	3.1	2.1	7.0	8.3	1.6	1.1	3.2	3.3

Note. The maximum equilibrium solubility, except for Al−Ti, is taken from data of [6].

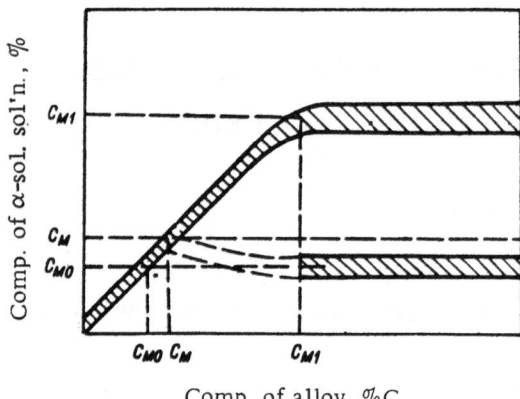

Fig. 1. Composition of primary solutions of carbon in Ni and Co formed at high cooling rates: C_M) maximum solubility according to the stable phase diagram; C_{M0} and C_{M1}) composition of solid solution obtained for cooling rates respectively below and above 10^5 deg/sec. For Ni−C alloys $C_M = 0.6$, $C_{M0} = 0.4$, and $C_{M1} = 1.8\%$ C, for Co−C alloys these are 0.8, 0.3, and 1.6% C respectively [2].

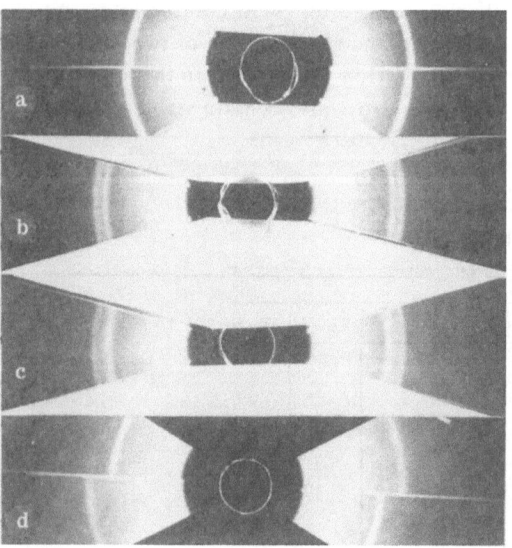

Fig. 2. X-ray photographs of Co−C and Ni−C alloys: 1) Ni−C solid solution C_{M0}; b) Ni standard; c) Ni−C solid solution C_{M1}; d) Ni−C solid solution $C_{M0} + C_{M1}$.

Solid solutions with intermediate compositions are not found in these conditions.

If the carbon content in the original melt is smaller than C_{M1}, and the cooling rate is sufficient for the formation of strongly saturated solid solutions, the formation of solid solutions of the same composition as the original liquid phase is possible.

In Al−Mn alloys, on increasing the cooling rate, three solid solutions with different solubilities are formed (Fig. 3).

Solid solutions within the equilibrium phase diagram (up to 1.4 wt. % Mn) are formed for cooling rates up to 10^2 deg/sec; others with 3 to 4% Mn occur for 10^2 to 10^4 deg/sec. Heavily supersaturated solid solutions containing 9.2 to 9.8% Mn only begin to crystallize for cooling rates exceeding 10^4 deg/sec. Further increasing the rate to 10^6 deg/sec produces no increase in solubility.

The crystallization of Al−Cu and Al−Mg alloys, which also have intermediate phases, was investigated, but the formation of these phases on solidification of the alloy occurred rather differently from those in Al−Mn alloys. Analysis of the solid solutions of alloys Al−Mg and Al−Cu showed the absence of any broadening in the solid-solution range, just as for alloys without intermediate phases, even for cooling rates of 10^6 deg/sec.

Discussion of Experimental Data

The solubility of one element in another is determined mainly by the relative dimensions of the atoms and the electron concentration [7]. However, when the system forms intermediate phases, the limiting solubility determined by the dimensions of the atoms and their structure is normally not observed. The precipitation of the intermediate phase stops the enrichment of the solid solution. On further raising of the concentration of the second component, it becomes thermodynamically more favorable to form a mixture of the solid solution of composition C_{M1} and the intermediate phase 1 (Fig. 4). If the less stable phase 2 crystallize instead of intermediate phase 1,

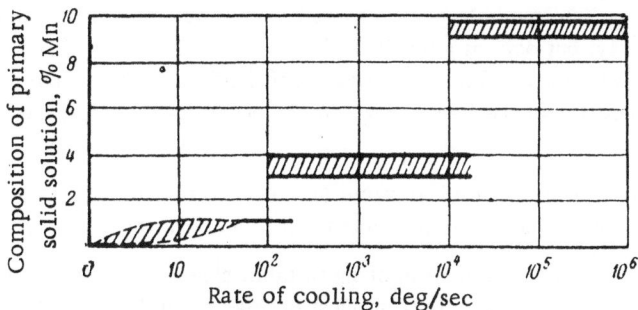

Fig. 3. Variation of the saturation of a solid solution of
Mn in Al with increasing cooling rate.

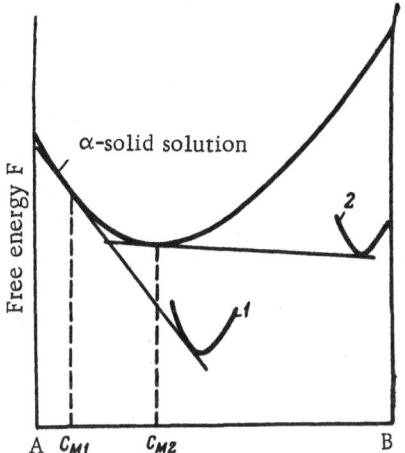

Fig. 4. Effect of the stability and composition
of the intermediate phase on the solubility of
primary solid solutions.

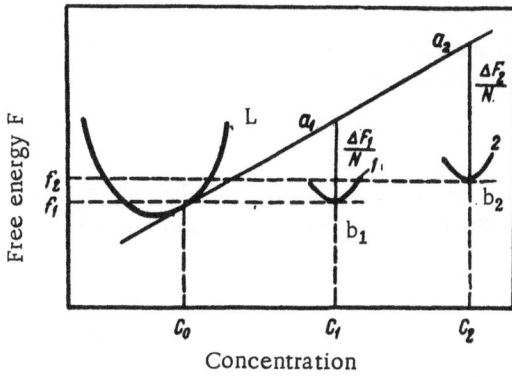

Fig. 5. Effect of the composition of the precipitating phase on the "motive force" of nucleus formation ($\Delta F/N$).

then the limiting solubility will be determined by the
concentration C_{M2} ($> C_{M1}$). This explains the discon-
tinuous character of the increase in the concentration of
the second component in the solid solutions on raising of
the cooling rate.

If in Ni−C and Co−C alloys the cooling rate is insufficient to suppress the precipitation of graphite, then
the limiting saturation of the primary solid solution will
correspond to concentration C_M to C_{M0} (Fig. 1). On crystallization of the metastable carbides, the solubility
jumps to C_{M1}.

An analogous picture is also found for crystallization of Al−Mn alloys.

A solubility up to 1.4 wt. % Mn lies within the limits of the stable phase diagram and corresponds to equilibrium of the solid solution with the most stable intermediate
phase $MnAl_6$. A solubility of 3 to 4% Mn is reached on
crystallization of the less stable phase $MnAl_4$ [8, 9]. The
sharp rise in solubility to 9.2 and 9.8% Mn is apparently
caused by the precipitation of the still less stable phase $MnAl_3$.

If the concentration of the original melt lies between C_M and C_{M1} (Fig. 1), and the cooling rate is sufficient for the formation of strongly supersaturated solid
solutions of composition C_{M1}, then the latter take the
composition of the original liquid. This is because the
motive force underlying the formation of the solid solutions is primarily the capacity of the elements for dissolution, the free energy of the solution being smaller
than that of any mixture of the original components. Thus
if the cooling rate is sufficient to supppress precipitation
of the stable phase (for example, phase 1 in Fig. 4), the
saturation of the solid solution extends further and may
reach the composition of the original liquid.

The precipitation of one intermediate phase or another does not, of course, affect the composition of the primary solid solutions directly, but only as a result of the effect on the composition of the liquid phase. Hence, strictly speaking, Fig. 4 should show the free-energy curve of the liquid phase. But since the final result is determined by the mutual disposition of the free-energy curves of the solid phases, the free-energy curve of the liquid phase is not given.

As shown above, heavily saturated solid solutions do not arise in all systems containing intermediate phases, evidently because the stable phases cannot be prevented from precipitating in all systems.

Let us consider the reason for the formation of metastable phases on crystallization of alloys at a high cooling rate. With respect to structure the alloys in question may be divided into two groups:

1. Ni−C, Co−C, Fe−C. In these alloys the metastable intermediate phases (carbides) are closer to the composition of the original liquid than the stable graphite. Hence crystallization of the carbides is associated with a smaller redistribution of concentration on the crystallization front, and may thus prove kinetically more favorable at high cooling rates [10].

2. Al−Mn, Al−Cr, Al−V, Al−Ti, Al−Fe, Sn−Sb. In these systems the metastable intermediate phases differ more from the composition of the original liquid than do the stable phases. Nevertheless, precipitation of the latter at high cooling rates may be inhibited [3, 6, 8, 9].

Analysis of the phase diagrams of the second group shows that all the phases existing in stable equilibrium with the primary solid solution are incongruent compounds and are formed by a peritectic reaction. The preparation of the liquid for crystallization of these compounds (in the sense of the development of the appropriate complexes of atoms) is very weak [12, 13]. Hence another phase develops in the melt at high cooling rates, and this, as experiment shows [6, 11], may continue its growth until the liquid is entirely used up. The stable phase will not precipitate here.

Of course, the precipitation of the less stable phase from the supersaturated matrix is not always the less thermodynamically favorable process. We see from Fig. 5 that the precipitation of the less stable phase 2 is associated with a greater change in the free energy per atom of the precipitating substance ($\Delta F/N$) than the precipitation of the more stable phase 1 [14, 15].

Analysis of the systems Al−Cu and Al−Mg shows that the stable phases (θ in Al−Cu and β in Al−Mg) are formed immediately from the liquid. Their melting is effected congruently. As shown by special investigations [12, 13], the melt in the region of such compounds contains complexes of atoms corresponding to the compound in the solid state. Thus the liquid is prepared for crystallization of the congruent compounds, and one cannot delay their crystallization even with very rapid cooling rates. Hence only stable phases crystallize from the melt at high cooling rates in Al−Cu and Al−Mg alloys, and the saturation of the primary solid solutions lies within the limits of stable equilibrium (α-solid solution+stable intermediate phase).

Conclusions

1. The broadening of the region of primary solid solutions on crystallization of alloys of the eutectic and peritectic types is feasible when the precipitation of the stable phase is eliminated and crystallization of a metastable intermediate phase becomes possible.

2. The precipitation of the stable phase on crystallization at a high cooling rate may be averted in two cases: When the phase is a noncongruent compound and is formed by a peritectic reaction, and when it differs more in composition from the original liquid than does the metastable phase.

3. For a correct description of the behavior of the system on crystallization of metastable phases, the metastable phase diagram must be used.

4. The rise in the solubility of primary solid solutions with increasing cooling rate takes place suddenly rather than gradually. A sharp increase in the content of the second component in the solid solution ensues when crystallization begins according to the metastable phase diagram.

5. The broadening of the region of solid solutions is not a direct consequence of the difficulty of diffusion in the liquid phase on crystallization at a high cooling rate. In addition to this, the fall in the mobility of the atoms may have an effect on the precipitation of the metastable intermediate phases and the suppression of the crystallization of the stable phases.

Literature Cited

1. I. S. Miroshnichenko and I. V. Salli, Zavodskaya Laboratoriya (11):(1959).
2. I. S. Miroshnichenko, Izv. Vuzov. Tsvetn. Metal. (1):(1961).
3. I. S. Miroshnichenko, Collection: "Crystallization and Phase Transformations" (Izd. Akad. Nauk Belorus. SSR, Minsk, 1962).
4. G. Falkenhugen and W. Hofmann, Z. Metallkunde 43(3):69 (1952).
5. N. N. Glagoleva, V. M. Glazov, and G. A. Korol'kov, Izv. Akad. Nauk SSSR, Otd. Tekhn. Nauk 8:89(1957).
6. M. Hansen and K. Anderko, Constitution of Binary Alloys [Russian translation] (Metallurgizdat, 1962).
7. W. Hume-Rothery and G. V. Raynor, Structure of Metals and Alloys [Russian translation] (Metallurgizdat, 1959).
8. H. Hanemann and A. Schrader, Atlas Metallographie, III, 2:119 (1952).
9. G. V. Raynor, J. Inst. Metals 11:507 (1944).
10. N. N. Sirota, Dokl. Akad. Nauk SSSR 64(5):697 (1949).
11. W. Geiler and H. Garbek, Arch. Eisenhüttenw. 10(26):611 (1955).
12. Collection: "Structure and Properties of Molten Metals" (Izd. A. A. Baikov Inst. Metallurgy, 1960).
13. V. M. Glazov, A. A. Vertman, and E. G. Shvidkovskii, Izv. Akad. Nauk SSSR, Metallurgiya i Toplivo (3):104 (1961).
14. H. K. Hardy, J. Inst. Metals 7:457 (1950).
15. A. H. Cottrell, Theoretical Structural Metallurgy [Russian translation] (Metallurgizdat, 1961).

FORMATION OF THE STRUCTURE OF EUTECTIC-TYPE ALLOYS
AT HIGH COOLING RATES

I. S. Miroshnichenko

The normal structure of eutectic-type alloys consists of primary precipitates of one of the components and the eutectic. However, as noted by Bochvar in 1935 [1], structural anomalies often occur in eutectic-type alloys:

1. At large supercoolings, alloys of noneutectic composition develop eutectic structures, which are usually called "quasieutectic" or "pseudoeutectic."
2. If a normal structure is formed in an alloy, then with increasing supercooling the amount of primary crystals falls and the amount of eutectic increases.
3. Depending on the conditions of crystallization, there may be a change in the structure of the eutectic itself.

Despite the fact that more than a quarter of a century has passed since Bochvar's work was published, the laws governing structure formation in systems of the eutectic type at various cooling rates are not very well known, especially for metal alloys.

There is no single view on the mechanism of the formation of quasieutectic structures.

Observing the crystallization of systems of transparent organic substances, Bochvar came to the conclusion that quasieutectic structures in systems of noneutectic composition arose when the linear crystallization rate of the eutectic exceeded that of the primary phase. The structure formation in this case proceeds by way of eutectic columns. The excess component settles out as a by-product, causing the eutectic laminas of the same composition to become thicker. It was also suggested that the region of quasieutectic systems was bounded by the continuation of the liquidus lines.

Popov and Shteinberg [2] in essence held to this view; they considered, however, that formation of the excess component would be completely suppressed, not at temperatures corresponding to the continuation of the liquidus on the phase diagram, but somewhat lower, at temperatures for which the rate of forming the eutectic (for the given composition of the liquid solution) exceeded that of forming the individual phases.

Grechnyi considered [3] that the formation of quasieutectic structures could not be due simply to the relative growth rates of the excess phase and the eutectic columns, and that the boundaries of the quasieutectic were not determined by extrapolating the liquidus. Grechnyi found the quasieutectic position on the phase diagram by simultaneously considering the continuation of the liquidus lines and the metastability boundaries of the two phases. According to this, the quasieutectic region could lie even outside the region bounded by the continuation of the liquidus lines.

Hardly any experimental attempts have been made to explain the effect of the degree of supercooling on the width of the quasieutectic region in alloys.

The formation of quasieutectic structures in $Fe-C$ [4] and $Al-Cu$ [5] alloys at high cooling rates is associated with substantial supercoolings, though the extent of the latter was not measured directly.

In [6], $Ni-Ni_3Sn$ alloys were supercooled by 95 to 178°C. No quasieutectic was obtained on crystallization, however, although the supercooling reached a region below the continuation of the liquidus lines.

Fig. 1. Microstructure of eutectics of slowly cooled alloys:
a) Fe−C (× 102); b) Al−Cu (× 60); c) Pb−Sn (× 72).

It is usually accepted [1-4] that the quasieutectic has the same morphology as the eutectic of the alloy. But since the morphology of the eutectic itself is affected by the crystallization conditions [1, 6], it is hard to see how the quasieutectic, the formation of which requires considerable supercooling, can always retain the morphology of the corresponding eutectic.

In the present investigation we studied the effect of the cooling rate on the structure formation of Fe−C, Al−Cu, and Pb−Sn alloys. The cooling-rate range was 0.5 to $5 \cdot 10^5$ deg/sec.

The following problems were posed: 1) to find the effect of the cooling rate on the structure of alloys with greater and smaller excesses of one of the components, and to determine the extent of the quasieutectic for cooling rates up to $5 \cdot 10^5$ deg/sec; 2) to study the morphology of the quasieutectic and decide whether it corresponded to that of the eutectic in the same alloys; 3) to study the effect of the cooling rate on the structure of the eutectic itself.

The alloys were obtained by fusing chemically pure materials in an electric furnace. The iron−carbon alloys were melted in vacuum and the Al−Cu and Pb−Sn in air, using protective fluxes.

A high rate of heat discharge was ensured by casting the alloys into metal chill molds, flattening out the drops of liquid metal with copper plates [7], and also catapulting the molten metal onto a copper slab. The cooling rate was calculated theoretically by Ryzhikov's method [8]. The final formula for determining the cooling rate took the following form:

$$V_{\text{cool}} = \frac{4b^2_{\text{m}} T^2_{\text{p}} (T_l - T_s)}{\gamma^2 S^2 \pi [L + C_l (T_l - T_s)]^2},$$

(1)

where

$$T_p = \frac{b_m}{b_m + b_l}\, T_m, \quad b_m = \sqrt{\lambda c \gamma}, \quad b_m = \sqrt{\lambda_m c_m \gamma_m}\,;$$

C_l is the specific heat of the metal, L is the heat of fusion, T_l is the temperature of the metal at the moment of casting, T_S is the temperature at the end of solidification (solidus temperature), $\lambda_{mo} c_{mo} \gamma_{mo}$ are the thermal conductivity, specific heat, and specific gravity of the mold material, λ, c, γ are the corresponding thermophysical coefficients of the crystallizing material, and S is the thickness of the crystallized layer.

For films formed by flattening out of the molten metal with copper plates, the cooling rate was also determined experimentally from the temperature change in the drop being flattened [7]. The cooling rates so found were close to those calculated from formula (1).

In all the alloys studied, at ordinary cooling rates, a structure with primary precipitates and a eutectic of lamellar form developed (Fig. 1). The lamellar form of the eutectic enabled changes in its morphology to be followed with ease.

In alloys with a large excess of one of the components, on increasing the cooling rate, structures with a distinct division of the eutectic phases appeared. The solidified alloys consisted of crystallites of the excess phase, in the intervals between which lay the monolithic second phase. The eutectic was not formed as an independent structural component (Fig. 2a and b). The capacity of a given alloy to crystallize with a distinct division between the eutectic phases increased with increases in the cooling rate, the quantity of excess phase, and the extent of its ramification.

In hypoeutectic alloys of the Fe−C system crystallized at a cooling rate of $5 \cdot 10^5$ deg/sec, a distinct phase separation is found for 2.2 to 3.4% carbon in the alloy. The Al−Cu alloys, for the same cooling rate, crystallize with a distinct separation of the phases if the copper in the alloy is less than 25 wt.%. In hypereutectic Fe−C and Al−Cu alloys it is considerably harder to suppress the formation of the eutectic as an independent structural constituent. Even in alloys with a large excess of the second component, eutectic columns develop together with the primary precipitates (Fig. 2c).

The formation of structures with a distinct separation of the eutectic components may be explained as follows.

On crystallization at a high cooling rate, considerable initial supercoolings of the original melt are obtained, as a result of which a large number of crystallization centers are formed. During the growth of crystals from the many centers, the gaps between them become so small that the deposition of one of the eutectic phases on the already forming primary crystals is more favored than the formation of new nuclei. An increase in the

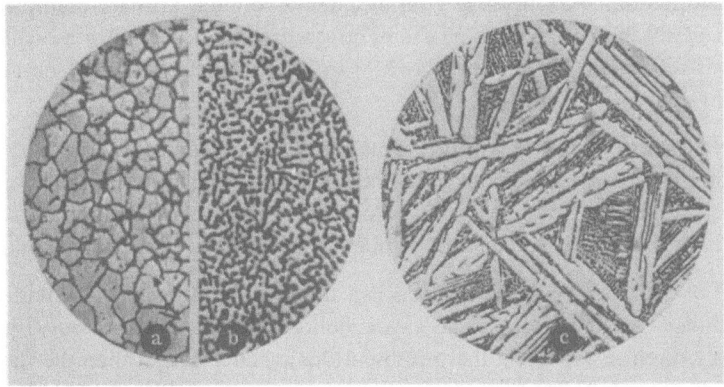

Fig. 2. Microstructure of rapidly cooled alloys: a) Fe−C, 2.2% C, etched by sodium picrate (× 900); b) Fe−C, 3.1% C, etched by sodium picrate (× 1200); c) Fe−C, 5.5% C, etched by 5% HNO₃ solution (× 540).

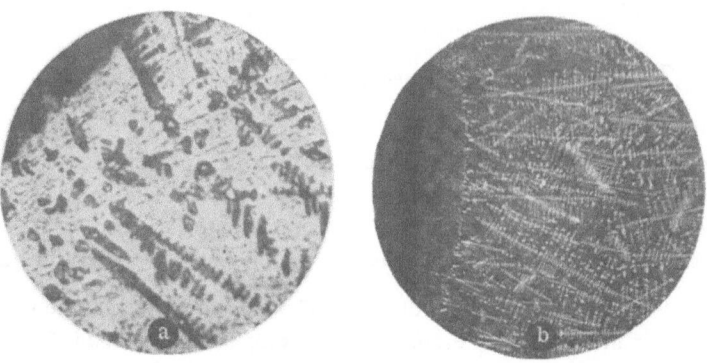

Fig. 3. Microstructure of the ends of Fe−C films (3.95% C):
a) etched in 5% HNO$_3$ solution (×1200); b) etched in sodium picrate (×360).

amount of the excess phase and the ramification of its precipitates facilitates the formation of structures with a
distinct phase separation. In hypereutectic Fe−C and Al−Cu alloys, where the primary phase crystallizes in the
form of poorly branched laminas, it is very hard to separate the phases constituting the eutectic.

In terms of tendency to form quasieutectics, the alloys studied may be placed in the following order:
Pb−Sn > Al−Cu > Fe−C.

In thin films (δ = 0.05-0.02 mm) of iron−carbon alloys, the quasieutectic is difficult to obtain even with
3.95% C in the alloy (Fig. 3a). In studying the ends of the film, one sometimes finds small regions adjacent to
the outer surfaces in which the formation of eutectic structures can be seen (Fig. 3b). It is hard to say, however,
without local analysis, whether these parts have the composition of the original liquid. The formation of quasieutectic structures in hypereutectic iron−carbon alloys has also not been observed. In alloys with 4.44% C,
primary cementite precipitates can be seen.

In Al−Cu alloys crystallized at cooling rates of $5 \cdot 10^5$ deg/sec, a quasieutectic appears in the concentration range 28 to 38 wt.% (Fig. 4).

The tendencies of hypo- and hypereutectic Pb−Sn alloys to form quasieutectic structures differ. Thus,
for example, whereas in hypereutectic alloys at cooling rates of 10^3 deg/sec the normal eutectic structure (β-
solid solution and eutectic) is not formed in the concentration range 61.9 to 90% Sn, in hypoeutectic alloys,
even close to the eutectic in composition, these cooling rates give clear primary precipitates of the α-phase.
With decreases in the heat-discharge rate, the quasieutectic region contracts. For a cooling rate of 294 deg/sec
the quasieutectic appears in the concentration range 61.9 to 75.0% Sn, for 225 deg/sec at 61.9 to 71.0%, and
for 110 deg/sec at 61.9 to 68.0% Sn.

If the quasieutectic is formed in one and the same sample together with the ordinary structure, the transition from one to the other is more or less sharp (Fig. 4c). A gradual fall in the amount of primary precipitates
on passing over to the quasieutectic is not observed. Moreover, raising the cooling rate of the crystallizing
alloy always leads to an increase in the amount of primary phase visible in the microsections.

Using the method of quantitative phase analysis [9], the relative amount of austenite in rapidly cooled
Fe−C alloys was determined (see the table). The experimental data were compared with theory. As seen in
the table, the amount of austenite visible in the microsections is much larger than the theoretical value calculated from the phase diagram [10] and the composition of austenite in thin films [11].

Analogous results are obtained on examining rapidly cooled Al−Cu alloys.

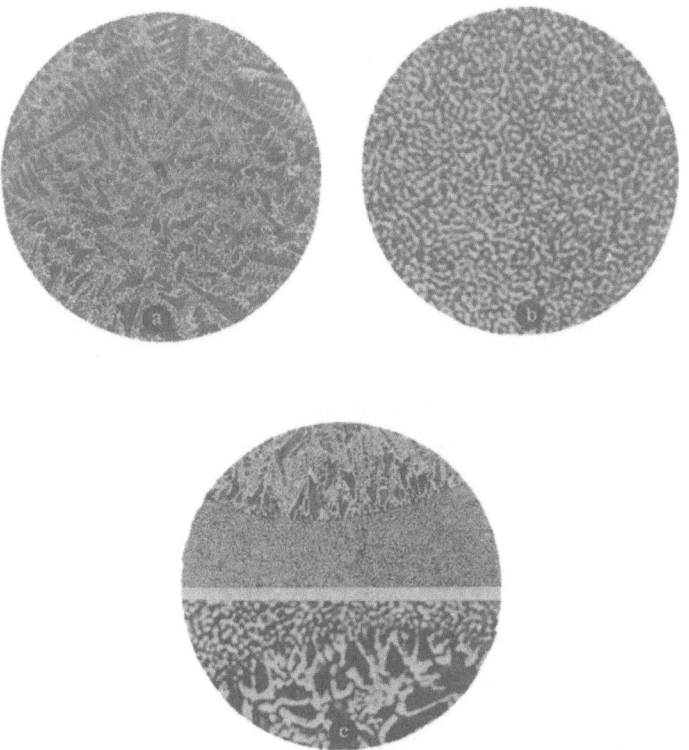

Fig. 4. Microstructure of rapidly cooled Al−Cu alloy (38 wt.% Cu): a) "normal" structure (primary precipitates and eutectic) (×780); b) quasieutectic (×1680); c) transformation from normal structure to quasieutectic (×510, ×1680).

The increase in the amount of primary phase visible in the microsections may be explained by the fact that, during the eutectic crystallization, there is a partial deposition of one of the eutectic phases on the earlier precipitated crystallites of the primary phase.

The more or less sharp transition from normal eutectic structure to the quasieutectic is apparently explained by the difference in the mechanisms by which these structures are formed. Confirmation of this is given by the fundamental difference in the morphology of the structures. Whereas the normal eutectic has a lamellar structure − a result of the simultaneous growth of the phases constituting the eutectic − the quasieutectic has a completely different form. This consists of grains of one phase surrounded by material of the second phase. The simultaneous growth of the two phases is excluded in obtaining such structures.

Amount of Austenite Visible in Microsections of Fe−C Alloys

% carbon in alloy	% carbon in austenite	Calculated amount of austenite, %	Actual amount of austenite, %	Number of experimental points	Absolute error
3.4	1.63	33.7	53.6	12600	0.26
3.9	1.35	13.5	28.6	14000	0.28

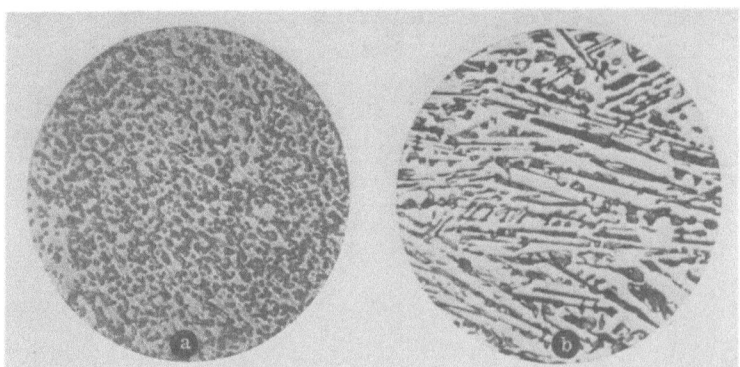

Fig. 5. Microstructure of the eutectic in rapidly cooled alloys:
a) Pb—Sn (× 900); b) Fe—C (× 900).

The transformation from the lamellar to the granular structure may also take place in alloys of eutectic composition. Thus, for example, in Pb—Sn alloys, even with cooling rates of 40 deg/sec, the eutectic structure normal for these alloys degenerates, loses its lamellar form, and acquires granular structure (Fig. 5a). In Fe—C and Al—Cu alloys, even at cooling rates of 10^5 to 10^6 deg/sec, the eutectic still retains its lamellar form (Fig. 5b), although in Al—Cu the lamellar nature is less clearly expressed.

The degeneration of the eutectic on crystallization at a high cooling rate may apparently be explained by the following.

With increasing supercoolings, the relation between the rates of formation and growth of the eutectic components changes. For some value of cooling rate, the probability of forming one phase becomes greater than that of the other. At the same time, the growth rate of the second phase is greater than that of the first. Thus the abundantly generated first-phase crystals cannot grow very extensively since the later-formed second phase, growing at a higher rate, surrounds them. Thus the simultaneous (cooperative) growth of eutectic columns is in this case converted into successive (interlaced) growth.

Since supercooling is required for the formation of the quasieutectic, we should naturally expect a transformation to granular structure here also, and for the same reasons. Moving away from eutectic composition leads to an increase in the probability of forming nuclei of the excess phase, which in its turn produces a transformation from simultaneous to interlaced growth of the structural components. Hence the formation of a granular quasieutectic is observed at lower cooling rates (and, correspondingly, supercoolings) than transformation from lamellar structure to granular in the eutectic of the same alloy. For this reason the morphology of the quasieutectic in general does not coincide with that of the eutectic obtained under the same conditions of crystallization.

In conclusion, we must note that we have only used the term "quasieutectic" because it has already been accepted in the literature. In fact, the author considers the choice of the expression "quasieutectic" to be rather regrettable. If we come to apply this term to a eutectic structure with a chemical composition differing from the eutectic [4], then in metal alloys at least this does not always occur. As shown above, the structure of the quasieutectic in general does not coincide with that of an alloy of eutectic composition obtained under the same conditions of crystallization. Furthermore, even the expression "eutectic structure," strictly speaking, does not make sense, since the latter depends on the conditions of crystallization.

Sometimes the quasieutectic is associated with a certain mechanism of crystallization, when there is simultaneous and cooperative growth of two solid phases in alloys of noneutectic composition [4]. The structures obtained in alloys of noneutectic composition at high cooling rates, however, may also not have the form which ought to result from cooperative growth. Moreover, the eutectic of these alloys, under the same conditions of crystallization, is formed as a result of cooperative growth.

Furthermore, the structures of the so-called "quasieutectics" in essence differ in no way from the structures with a distinct separation of eutectic phases which we obtained in alloys with a large excess of one of the components (cf., Figs. 4b and 2b). Moreover, the latter were formed for supercoolings at which the temperature-concentration region for the formation of quasieutectic structures (according to [1, 2]) had not been reached.

Conclusions

1. Alloys of the eutectic type containing a predominance of one of the components crystallize with a distinct separation of the eutectic phases at high cooling rates. The structure so formed consists of grains of the excess phase in the interstices between which lies the monolithic second phase. The eutectic is not formed as an independent structural component. The capacity of the alloy to crystallize with a distinct separation of phases increases with cooling rate and amount of excess phase.

2. Whereas at high cooling rates the normal structure (primary precipitates and eutectic) forms in the alloy, the amount of primary precipitates visible on the microsections is greater than it should be according to the equilibrium phase diagram.

3. In alloys close to eutectic composition, a quasieutectic with a structure in general differing from that of the eutectic obtained under the same solidification conditions may crystallize.

In our alloys, the quasieutectic took on a granular structure, indicating a successive (interlaced) growth mechanism of the components in the quasieutectic.

4. At high cooling rates, there is a change in the morphology of the eutectic in Pb−Sn alloys. These transform from the lamellar structure to the granular. At cooling rates of 10^5 to 10^6 deg/sec, the sharp lamellar structure in the Al−Cu eutectic also begins to disppear.

5. The formation of a granular quasieutectic occurs at lower cooling rates than the transition from lamellar to granular structure in the eutectic of the same alloy.

Literature Cited

1. A. A. Bochvar, Study of the Mechanism and Kinetics of Crystallization in Alloys of the Eutectic Type (ONTI, 1935).
2. S. S. Shteinberg, Physical Metallurgy (Metallurgizdat, 1961).
3. Ya. V. Grechnyi, Izv. Akad. Nauk SSSR, Otd. Tekhn. Nauk (3):77 (1956).
4. K. P Bunin and Ya. N. Malinochka, Introduction to Metallography (Metallurgizdat, 1954).
5. E. Scheil and I. Masuda, Aluminum 2:51 (1955).
6. W. Geller and H. Garbeck, Arch. Eisenhüttenw., 10:611 (1955).
7. I. S. Miroshnichenko and I. V. Salli, Zavodskaya Laboratoriya 11:1398 (1959).
8. A. A. Ryzhikov, Theoretical Bases of Foundry Production (Mashgiz, 1954).
9. S. A. Saltykov, Stereometric Metallography (Metallurgizdat, 1958).
10. A. Hansen and K. Anderko, Constitution of Binary Alloys [Russian translation] (Metallurgizdat, 1962).
11. I. S. Miroshnichenko, Collection: "Crystallization and Phase Transformations" (Izd. Akad. Nauk Belorus. SSSR, Minsk, 1962).

KINETIC EQUATIONS OF ALLOY CRYSTALLIZATION

V. T. Borisov

A phase transformation taking place at a constant rate constitutes an irreversible phenomenon. Application of the thermodynamic theory of irreversible processes [1] enables us to find the general form of the kinetic equations and study the role of thermodynamic factors characteristic of the transformation under consideration. This approach gives results suitable for describing any phase transformation of the first kind, but we shall be considering the specific case of the crystallization of a multicomponent system.

Let us consider a system consisting of two phases (') (liquid) and (") (solid alloy) separated by a plane boundary 1 cm^2 in area moving at velocity V (Fig. 1). Let the temperature T, pressure P, and chemical potential μ_k of the k-th component have the same values at all points of the volume of each phase. The expression of the second law of thermodynamics, written for any phase in the form

$$T dS = dU + P dv - \sum_{k=1}^{n} \mu_k dN_k,$$

allows for the possibility of a change in entropy S as a result of changes in energy U, volume v, and number N_k of particles of the k-th kind. The change in the entropy of the whole system takes the form

$$dS = \frac{d_2 U' + P dv'}{T'} + \frac{d_2 U'' + P dv''}{T''} + \frac{d_1 U'}{T'} + \frac{d_2 U''}{T''} -$$
$$- \sum_{k=1}^{n} \left(\frac{\mu_k'}{T'} dN_k' + \frac{\mu_k''}{T''} dN_k'' \right).$$

$$(1)$$

In contrast to the pressure P' = P" = P (mechanical equilibrium), the temperatures T' and T" cannot in advance be considered identical, since at the phase boundary the transformation proceeds at a finite rate. The change in energy of each phase in formula (1) is calculated for the internal and external parts dU = $d_1 U$ + $d_2 U$ so that dU = $d_1 U$ represents the change in energy of the given phase as a result of transport from the adjacent phase, and $d_2 U$ as a result of interaction with the external medium. The first two terms of (1), connected with the transfer of certain quantities of heat from the surrounding medium (dq' = $d_2 U'$ + Pdv', dq" = $d_2 U''$ + Pdv"), constitute the reversible part of the entropy change, which is conserved under quasi-steady-state conditions, and the rest characterize the strictly irreversible process. Considering that $d_1 U'$ + $d_1 U''$ = 0, in view of the law of energy conservation, and dN'$_k$ + dN"$_k$ = 0, if exchange by the particles occurs only between different phases, we obtain for the rate of increase of the irreversible part of the entropy the expression

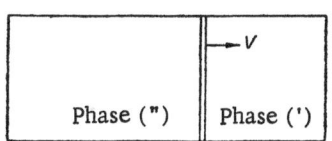

Fig. 1. Scheme for deriving kinetic equations for the growth of a phase.

$$\frac{dS}{dt} = -\frac{\Delta T}{T^2} \frac{d_1 U''}{dt} + \sum_{k=1}^{n} \left(\frac{\mu_k' - \mu_k''}{T} + \frac{\varepsilon_k''}{T^2} \Delta T \right) \frac{dN_k''}{dt}.$$

$$(2)$$

Here the values of functions μ_k and ε_k (energy belonging to an atom of the k-th component) are referred to the same temperature $T = T'$ and the notation $\Delta T = T'' - T'$ is introduced. The entropy–balance equation could be taken directly from general theory [1]. Its derivation is briefly reproduced above so as to emphasize that it remains valid also in the case where a phase transformation is taking place and the phase-separation boundary is displaced.

Relation (2) enables us to determine the conjugate flows and forces in the following way:

$$I_k = \frac{dN_k''}{dt}, \quad I_0 = \frac{d_1 U''}{dt},$$

$$X_k = \frac{\mu_k' - \mu_k''}{T} + \frac{\varepsilon_k''}{T^2} \Delta T, \; X_0 = -\frac{\Delta T}{T^2},$$

where I_k is the number of atoms of type k $(K = 1, 2, \ldots, n)$ passing from the liquid to the solid phase through 1 cm^2 of the surface of separation per unit time, I_0 being the energy flux in the same direction. The kinetic equation here takes the form

$$I_i = \sum_{j=0}^{n} \alpha_{ij} X_j, \; i = 0, 1, \ldots, n,$$

and Onsager's relation $\alpha_{ij} = \alpha_{ji}$ is satisfied.

The most important case of the crystallization of an alloy is the steady-state condition. This takes place if the displacement rate V of the surface of separation is kept constant for a fairly long time. If the relaxation time of the processes accompanying the crystallization of the alloy is fairly small, the condition may be regarded as steady-state even in the case where it does not appear so on the macroscopic scale. For constant velocity V, the fluxes I_k $(k = 1, 2, \ldots, n)$ connected with the displacement of masses are also constant. The energy flux I_0, under these limitations, should assume a value corresponding to a minimum entropy growth rate. The latter is represented by the positive-definite form

$$dS/dt = \sum_{i=0}^{n} X_i I_i.$$

Carrying out the necessary operations

$$0 = d\frac{dS}{dt} = \sum_{i=0}^{n} I_i dX_i + X_0 dI_0 \; = \sum_{j=0}^{n} X_j \sum_{i=0}^{n} \alpha_{ji} dX_i + X_0 dI_0 = 2X_0 dI_0$$

(we have used the condition for constancy of the fluxes

$$\sum_{i=0}^{n} \alpha_{ji} dX_i = 0$$

with i = 1, 2, ..., n and symmetry of the coefficients α_{ij}), we find that the steady state is characterized by the vanishing of the force X_0 (if the matrix $\| \alpha_{ij} \|$ is not degenerate), i.e , the interacting phases have the same temperature in the steady condition of crystallization or melting. The kinetic equations of the steady process thus express the following association between the particle fluxes and the chemical-potential differences:

$$I_k = \sum_{j=1}^{n} \frac{\alpha_{kj}}{T}\,(\mu_j' - \mu_j''),\ \ k = 1, 2, \ldots, n$$

(3)

and give an expression for the transfer of energy from one phase to the other:

$$I_0 = \sum_{k=1}^{n} \alpha_k I_k,$$

(4)

where

$$\alpha_k = \sum_{j=1}^{n} \alpha_{j0}\alpha_{kj}^{*}\ \ \text{and}\ \ \|\,\alpha_{kj}^{*}\,\|$$

is the reciprocal matrix.

The rate of displacement of the phase boundary may be introduced by means of the relation, $I_k = VC_k''/\Omega$, where C_k'' is the concentration of the k-th component in molar proportions and Ω is the volume for one particle in the solid phase. Let us write out the equation for the case when cross phenomena may be neglected. The energy flux here vanishes, and expressions (3) are written in the form

$$C_k'' V = B_k C_k'\,(\mu_k' - \mu_k''),$$

(5)

where B_k are the new kinetic coefficients ($B_k = \Omega\alpha_{kk}/TC_k'$).

Equation (5), or in the general case (3), determines the growth rate and composition of the solid phase crystallizing from a liquid of given composition at a given supercooling. For a binary system, Eqs. (4) were analyzed on the regular-solution approximation in [2]. The solution of system (5) (k = 1, 2) may be conveniently represented in the form of a set of kinetic diagrams, having in a certain respect the geometrical properties of equilibrium phase diagrams. The extent of the deviation of the kinetic diagrams from the equilibrium ones depends primarily on the absolute values of all the kinetic coefficients for the given system.

If, for example, Eq. (5) is written down for an ideal binary system (k = a, b; $\mu_k = \mu_{k0} + RT\ln C_k$), confining ourselves to a linear expansion of the potentials in the neighborhood of the equilibrium state, we obtain the expressions

$$\frac{q}{T_l}\,\Delta T = \left(\frac{p_b C_a''}{B_a} + \frac{p_a C_b''}{B_b}\right) V,\ \ \frac{qRT_l}{C_a'' C_b''}\,\Delta C_b'' = \left(\frac{p_a q_a}{B_b} - \frac{p_b q_b}{B_a}\right) V,$$

(6)

determining the deviation of the temperature ΔT and concentration $\Delta C_b''$ of the solid phase precipitating from liquid of composition C_a' from the corresponding equilibrium values for a given growth rate. The notation is indicated in Fig. 2, apart from which $p_a = C_b''/C_b'$, $p_b = C_a''/C_a'$, $q = q_a C_a'' + q_b C_b''$, where q_a, q_b are the heats of fusion of the pure substances. As we see, the temperature change is always positive, while the composition of the solid phase may deviate either toward enrichment or toward impoverishment of the dissolved component. It is possible that the change in composition of primary crystals for very high cooling rates of the system observed in [3] is connected with the phenomenon discussed.

For a series of pure metals, the kinetic coefficients have quite high values [4], and deviations from the melting point are not directly manifested. This is probably a characteristic property of the majority of metals.

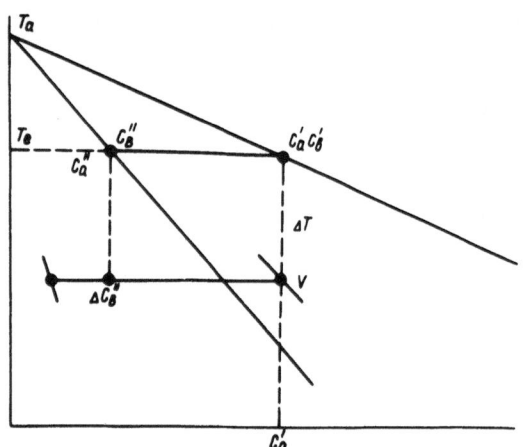

Fig. 2. Displacements of temperature and concentration of the solid phase with respect to their equilibrium values on crystallization of an alloy with finite velocity V.

Available experimental data will not yet allow us to estimate the quantities B_k for metal alloys directly, but certain facts apparently indicate that, on passing from a pure substance to an alloy, there is a considerable deviation of the temperature at the crystallization front from the temperature of the liquidus [5]. According to [6], this may be connected with the fact that one of the two kinetic coefficients is fairly small. In the neighborhood of the a corner of the phase diagram, the quantity B_b tends rather to be small, while B_a remains close to its value for the pure substance.

The growth of a phase in a multicomponent system was considered kinetically in [6]. On the basis of general concepts of the theory of reaction rates, expressions were obtained for the fluxes j'_k of k-atoms from the liquid phase into the solid and j''_k in the reverse direction:

$$j'_k = \beta'_k v'_k C'_k e^{-Q'_k/RT}, \quad j''_k = \beta''_k v''_k C''_k e^{-Q''_k/RT}, \tag{7}$$

where Q'_k, Q''_k are the activation energies for the transformation of atoms from one phase into the other in the corresponding directions, v'_k, v''_k are the vibration frequencies, and β'_k, β''_k may be called accommodation coefficients, characterizing the probability of activated atoms of the k-th sort joining the corresponding phase. The resultant flux is $I_k = j'_k - j''_k$, and if we introduce the growth rate V of the phase, then

$$C''_k V = \Omega j'_k \left(1 - \frac{\beta''_k v''_k C''_k}{\beta'_k v'_k C'_k} e^{-\frac{Q''_k - Q'_k}{RT}} \right). \tag{8}$$

Some of the quantities in this expression may be expressed in terms of thermodynamic characteristics if we use the proposition [7] that the chemical potential of the k-th component of a real solution may to a high degree of approximation be written as the sum

$$\mu_k = \mu_{k0} + RT \ln C_k + \varepsilon_k, \tag{9}$$

where the first two terms express the chemical potential of the component in an ideal solution, and ε_k is the excess potential energy belonging to one atom of the k-th sort in the actual system as compared with the ideal. Expression (9) is equivalent to the assumption that the activity coefficient depends in an explicit way only on the potential energy of the atoms of the given sort; it is satisfied for ideal and regular solutions, as may be readily confirmed. The region of applicability of this expression is determined by the fact that the difference between approximation (9) and the exact expression is due solely to the difference in the entropy terms corresponding to the actual and the regular alloy, since the energy term is written down without any assumptions.

For the difference between the potentials in the two phases we obtain, using the form $\mu_{k0} = \varepsilon_{k0} - Ts_{k0}$,

$$\mu'_k - \mu''_k = (\varepsilon'_{k0} + \varepsilon'_k) - (\varepsilon''_{k0} + \varepsilon'_k) - T(s'_{k0} - s''_{k0}) - RT \ln(C''_k/C'_k).$$

Assuming as before that the properties of each of the phases are conserved right up to the separation boundary, the difference between the first two brackets must be equated to the activation-energy difference $Q''_k - Q'_k$. Remembering also that $S'_{k0} - S''_{k0} = q_k/T_k$, where T_k is the melting point of the pure component, we obtain the relations

$$\frac{C''_k}{C'_k} e^{-\frac{Q''_k - Q'_k}{RT}} = e^{-\frac{q_k}{RT_k}} e^{-\frac{\mu'_k - \mu''_k}{RT}},$$

(10)

with the aid of which the kinetic equations arrive at the form

$$C''_k V = \Omega j'_k \left(1 - \frac{\beta''_k \nu''_k}{\beta'_k \nu'_k} e^{-\frac{q_k}{RT_k}} e^{-\frac{\mu'_k - \mu''_k}{RT}} \right).$$

(11)

For equilibrium of the phases, the difference $\mu'_k - \mu''_k$ and growth rate V should vanish simultaneously. A necessary condition for this is the following relationship between the accommodation coefficients of k-type atoms of the solid and liquid phases:

$$\beta'_k = \beta''_k \frac{\nu''_k}{\nu'_k} e^{-q_k/RT_k}.$$

(12)

The difference in (11) becomes equal to $1 - \exp(-\Delta\mu_k/RT)$ and for slight deviation from equilibrium depends linearly on $\Delta\mu_k$. Thus

$$C''_k V = \frac{\Omega j'_k}{RT} (\mu'_k - \mu''_k) = B_k C'_k (\mu'_k - \mu''_k),$$

(13)

i.e., the kinetic and thermodynamic considerations of crystallization kinetics in the first approximation lead to the same equations. According to (13) and (7), the kinetic coefficients have the structure

$$B_k = \frac{\Omega \beta'_k \nu'_k}{RT} e^{-Q'_k/RT}$$

(14)

and do not depend explicitly on the composition of the alloy, which justifies the form of description in Eqs. (5).

Literature Cited

1. S. R. de Groot, Thermodynamics of Irreversible Processes [Russian translation] (GITTL, Moscow, 1956).
2. V. T. Borisov, Dokl. Akad. Nauk SSSR 142(1):69 (1962).
3. I. S. Miroshnichenko, Present collection, pp. 55, 61.
4. V. T. Borisov and A. I. Dukhin, Probl. Metalloved. i Fiz. Met., Vol. 7 (Metallurgizdat, 1962), pp. 363-75.
5. V. T. Borisov, A. I. Dukhin, Yu. E. Matveev, and É. P. Rakhmanova, Present collection, p. 75.
6. K. A. Jackson, Can. J. Phys. 36(6):683 (1958).
7. V. K Semenchenko, Surface Phenomena in Metals and Alloys (GITTL, Moscow, 1957).

EXPERIMENTAL DETERMINATION OF KINETIC COEFFICIENTS
FOR BINARY SYSTEMS

V. T. Borisov, A. I. Dunkhin, Yu. E. Matveev, and É. P. Rakhmanova

The kinetic equations for a binary system with approximately regular solutions have the form [1, 2]

$$V(1-y) = B_a(1-x)\left[x^2 U' - y^2 U'' + RT \ \ln \frac{1-x}{1-y} + q_a\left(1 - \frac{T}{T_a}\right)\right],$$

$$Vy = B_b x\left[(1-x)^2 U' - (1-y)^2 U'' + RT \ \ln \frac{x}{y} + q_b\left(1 - \frac{T}{T_b}\right)\right].$$

$$(1)$$

Here V is the displacement rate of the phase boundary, q_a and T_a are the heat of fusion and melting point of component a, B_a is the corresponding kinetic coefficient, x and U' are respectively the concentration of the b component and the energy of mixing in the liquid phase, and y and U" are the concentration of component b and the energy of mixing of the solid phase.

In the absence of solubility in the solid phase, only one of the two equations (1) remains. Considering that $y \rightarrow 0$, we obtain

$$\frac{V}{B_a(1-x)} = x^2 U' + RT \ln(1-x) + q_a\left(1 - \frac{T}{T_a}\right).$$

$$(2)$$

Subtracting the equation of equilibrium (T_l is the temperature of the liquidus for a system with a concentration x of the b component),

$$0 = x^2 U' + RT_l \ln(1-x) + q_a\left(1 - \frac{T_l}{T_a}\right).$$

$$(3)$$

and supposing that the energy of mixing does not depend on temperature, we find an explicit expression for the growth rate of primary crystals of a:

$$V = B_a R\left[\frac{q_a}{RT_a} - \ln(1-x)\right](1-x)\Delta T,$$

$$(4)$$

where the supercooling ΔT is reckoned from the liquidus temperature of the system. From experimental values of V and ΔT we can determine the value of the single (in the present case) kinetic coefficient B_a.

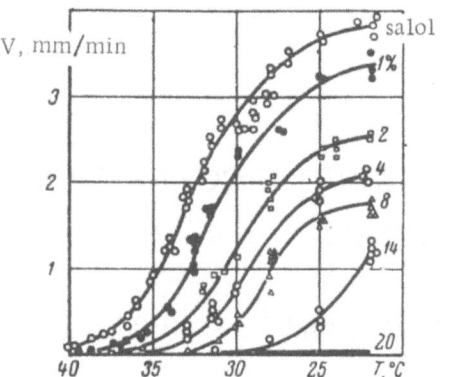

Fig. 1. Variation of the growth rate of salol crystals in the salol—azobenzene system with temperature.

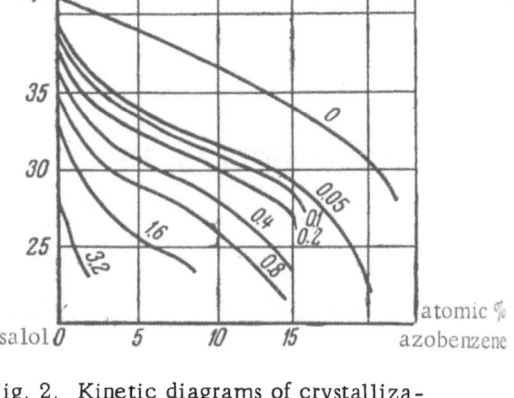

Fig. 2. Kinetic diagrams of crystallization for the salol—azobenzene system, corresponding to various growth rates (rates given in mm/min).

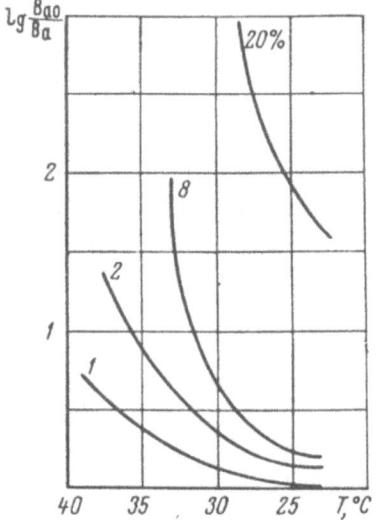

Fig. 3. Variation of the kinetic-growth coefficient for salol with composition and temperature.

Experiments were made on the salol—azobenzene system. The experimental technique normal for such cases was principally used [3]. Into a fine layer of melt lying between two glasses and kept in a thermostat was introduced a seed of pure salol, and the rate of displacement of a single sharp point of the crystal was measured in the microscope (in all cases, including that of pure salol, the same sharp corner of the crystal was selected, so that the orientation of the crystal relative to the growth direction remained the same). The results of such measurements for melts of different compositions are shown in Fig. 1. In Fig. 2 the same data are presented in the form of kinetic diagrams, from the shape of which we may qualitatively judge the strong dependence of the kinetic coefficient for salol on the concentration of azobenzene.

In a number of cases the experiments were carried out differently. A system of given concentration was held for 1 to 2 days at a temperature very close to that of the liquidus, so that equilibrium was established in the melt between the liquid and a small number of individual salol crystals (according to our estimates, the solubility of azobenzene in solid salol was insignificant). Differences in concentration between points of the melt at different distances from the crystallites were also thus leveled out. After this the specimen was rapidly transferred to a second thermostat, the earlier-established temperature of which ensured the required supercooling. A graph relating the position of the vertex of the growing crystal to time was then drawn which gave the growth rate at any moment. The growth rate fell with time owing to the enrichment of the region near the growing crystal with azobenzene. The rate at the first instant of time was determined by extrapolation; this may be related to the original concentration of azobenzene in the liquid (known from the preparation conditions and the equilibrium temperature) with greater confidence than in the case of the experiments described above. It proved, however, that the values obtained by different methods differed little from one another, so that the difference lay within the limits of experimental error. This is apparently due to the fact that the salol crystals have a tip with an edge lying in the plane of the glasses bounding the layer, so that the azobenzene is repelled not only into the region in front of the growing crystal but also toward the glass surfaces above and below.

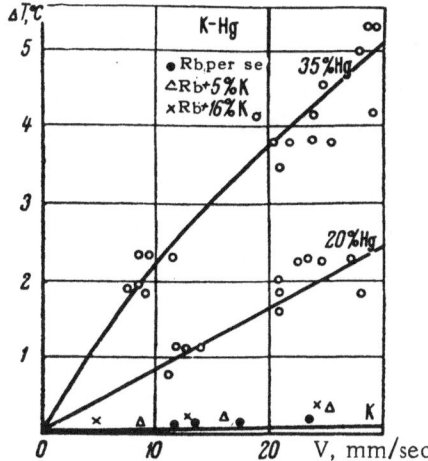

Fig. 4. Variation of supercooling at the beginning of the two-phase zone with growth rate for alloys.

At each temperature T, the growth rate V_0 of crystals in pure salol was also measured. From (4) we obtain $V_0 = B_{a0}q_a(T'_a - T)/T_a$, so that the relation

$$\frac{B_{a0}}{B_a} = \frac{V_0}{V}\left[1 - \frac{RT_a}{q_a}\ln(1-x)\right]\frac{T_l - T}{T_a - T}(1-x)$$

(5)

enables us to estimate the ratio of the kinetic growth coefficients for salol in the pure substance and in the binary system.

Figure 3 shows the ratio of these coefficients as a function of concentration and temperature. We notice the sharp fall in the kinetic coefficient B_a on increasing the concentration of the added substance. Clearly this cannot be explained by the dependence of such quantities as mobility or viscosity on composition, since the data of Fig. 3 characterize the behavior of the ratio B_{a0}/B_a. This dependence is evidently connected with characteristics of the molecular growth mechanism, in the present case with the kinetics of the generation and growth of two-dimensional nuclei on the faces of the crystals.

Qualitatively similar laws were found on studying metal systems (alloys). Figure 4 shows the results of an investigation into the K−Hg system. In the pure metal, supercooling at the crystallization front is absent even for growth rates on an order of 30 mm/sec. The deviation of the straight line constructed from the data of [4] for pure potassium from the axis of abscissas only characterizes experimental error. In potassium−mercury alloys, however, there is a clear dependence of the supercooling (reckoned from the alloy−liquid temperature) on the concentration and growth rate. The experiments were conducted according to the method of [4], using small drops, the initial temperature of which was always above the temperature of the liquidus.

On crystallization of the alloy, the separation boundary between the liquid and solid phases was severely broken up, so that it was more reasonable to speak of a transitional two-phase zone rather than the sharp crystallization front characterizing the pure metal, when the crystal grows in a nonsupercooled melt. The two-phase zone almost entirely eliminates the so-called diffusion supercooling in the liquid, which would have taken place if the separation boundary had been plane [5], but it is nevertheless possible that partial supercooling of the liquid remains. In this case the supercooling shown for alloys in Fig. 4 may be connected not only with a deviation of the temperature at the phase-separation boundary from the equilibrium value, but also with residual diffusion supercooling of the liquid within and in front of the two-phase zone.

However, two circumstances indicate that the graphs shown in Fig. 4 are associated with something more than single diffusion supercooling: the approximately linear dependence on growth rate (especially for the alloy with 20 wt.% mercury) characteristic of supercooling at the phase boundary, and the series of experiments made for rubidium − potassium alloys. The points shown on Fig. 4 indicate that in this case there is no dependence of ΔT on the composition of the alloy over quite such a wide range of growth rates, despite the fact that the temperature interval between the liquidus and solidus lines (1 to 2°C for an alloy with 5 wt.% K) represents quite a large distance on the scale of the figure. If we assume that this fact indicates a relatively small diffusion-supercooling effect in the two-phase zone, then the absence of any dependence of ΔT on growth rate may be explained by the large kinetic coefficients for RB, K, and their alloys, where, in contrast to the K−Hg system, the coefficients B_a and B_b depend very little on composition, owing to the similarity between the physical and chemical properties of the elements.

Literature Cited

1.   V. T. Borisov, Present collection, p. 69.
2.   V. T. Borisov, Dokl. Akad. Nauk SSSR 142(1):69-71 (1962).
3.   A. A. Bochvar, Study of the Mechanism and Kinetics of the Crystallization of Eutectic-Type Systems (ONTI, 1935).
4.   V. T. Borisov and A. I. Dukhin, Probl. Metalloved. i Fiz. Met., No. 7 (Metallurgizdat, 1962), pp. 363-74.
5.   G. P. Ivantsov, "Diffusion" Supercooling on Crystallization of a Binary Alloy, Dokl. Akad. Nauk SSSR 81(2):179 (1951).

EFFECT OF THE MORPHOLOGY OF THE ETCH FIGURES ON THE FORM ASSUMED BY METAL CRYSTALS DURING DISSOLUTION

I. M. Novosel'skii

The dislocation theory of crystals provides the best picture of the dissolution process. The essence of this picture lies in the fact that the dissolution of a crystal is strongly localized at the sites of etch figures, the formation of which is closely linked with the presence of dislocations. The question of the mechanism of the formation of etch pits has not been particularly elucidated in the literature, although it is considered that there are few ways in which dislocations develop into etch pits. Thus a screw dislocation on dissolution can give three or four planes which will face an etch pit [1]. Experimental study of the morphology of etch figures in single crystals of face-centered cubic metals shows that in almost all cases the etch figures are bounded by (100) or (111) planes or their combinations [2-10]. The effect of the external medium and conditions of dissolution reduces to the acceleration of some such possibilities and the retardation of others. An important part in this is played by the adsorption of anions, cations, or other particles at the crystal face and by the steric correspondence factor between these and the metal atoms at the crystal surface [2, 11-13].

Correlation of a large amount of experimental data on the anisotropy of the dissolution rates of metal crystals and the etch figures resulting therefrom shows that the anisotropy of the dissolution rates and electrode potentials of the faces is determined by the morphology of the etch figures [14].

The anisotropy of properties arising in the dissolution of crystals (dissolution rate, electrode potentials, etc.) is usually considered from the standpoint of thermodynamic considerations only [15-20], according to which the anisotropy of the properties of the dissolving crystal is determined only by the properties of the crystal itself (different packing densities of the metal atoms and ions on the crystal faces, anisotropic electron-density distribution). A review of the literature on the anisotropy of the properties of dissolving crystals shows that this is almost always inherent to the crystals, but may appear differently under specific conditions of dissolution (see the table).

It may be seen from the table that to each variation in the order of the dissolution-rate anisotropy there corresponds a change in the morphology of the etch pits. In aluminum this is observed on changing acid solutions into alkaline; in copper, changes occur upon the replacement of ammonium persulfate solution by nitric acid or acid solutions of copper sulfate and chloride. The dissolution rate is always smaller in the direction of crystallographic indices, which coincide with those of the planes bounding the etch figures.

Microscopic and electron-microscopic examination of crystal surfaces, after prolonged dissolution (for example, in [4-6, 28]) very often shows quite definite and sharp etch figures characterizing each layer. This proves that we must not regard the etch figures as the first stage in the dissolution of the crystals [29]. The etch figures arising at the onset of dissolution exist during the whole process of crystal decomposition, until the crystal has completely dissolved. If we consider that the figures are the result of directional decrystallization of dislocations or other imperfections, and that their shape appears as a result of elementary acts of interaction between the crystal surfaces and the external medium, then we may say that the etch figures "conduct" the dissolution process.

It is sufficient to know the morphology of the etch pits in order to explain or predict the experimentally observed order of the dissolution rates of various crystal faces. The relations between the dissolution rates and

Morphology of Etch Figures and Anisotropy of Dissolution Rates

Metal	Composition of solution	Etch-figure boundary	Reference	Composition of solution	Series of falling dissolution rates	Reference
Al	HCl conc., 1h; H_2O, 2h	(100)	21	HCl conc.	$V_{[111]} > V_{[100]}$	22
	HCl, 10%	(100)	2	HCl 3 N	$V_{[111]} > V_{[110]} > V_{[100]}$	23
	HCl from 0 to 6 M in dioxane	(100)	3			
	HCl > 6 M in dioxane	(111)	3	—	—	
	HBr, 40%	(100)	2			
	HI, 10%	(100, 111)	2			
	HCl, sp. gr. 1.17, 2 h	(100)	4, 5, 6	HCl, sp. gr. 1.17, 2h	$V_{[111]} > V_{[110]} > V_{[100]}$	4.5
	HNO_3, sp. gr. 1.41, 1 h			HNO_3, sp. gr. 1.41, 1h		
	HF, 48%, 0.2 h			HF, 48%, 0.2 h		
	NaOH, 15%	(335)	4	NaOH, 15%	$V_{[100]} > V_{[110]} > V_{[111]}$	4
Ag	NH_4OH, 5%	(100)	7	NH_4OH, 5%	$V_{[111]} > V_{[110]} > V_{[100]}$	7
	H_2O_2, 3%			H_2O_2, 3%		
Cu	$CuSO_4$, 205 g/liter	(100)	24	$CuSO_4$, 205 g/liter	$V_{[111]} > V_{[100]}$	24
	H_2SO_4, 48 g/liter			H_2SO_4, 48 g/liter		
	$CuCl_2$ + HCl	(100)	8	HCl, 0.3 N	$V_{[111]} > V_{[110]} > V_{[100]}$	25
				H_2O_2, 0.1 N		
	$(NH_4)_2S_2O_8$, 300 g/liter	(111)	8	$(NH_4)_2S_2O_8$, 02 N	$V_{[110]} > V_{[100]} > V_{[111]}$	25
	HNO_3, 20%	(100)	7	HNO_3, 20%	$V_{[111]} > V_{[110]} > V_{[100]}$	7
Pb	HNO_3, 20%	(100)	7	HNO_3, 20%	$V_{[111]} > V_{[110]} > V_{[100]}$	7
	CH_3COOH, 3%	(100)	7	CH_3COOH, 3%	$V_{[111]} > V_{[110]} > V_{[100]}$	7
	HNO_3, 4%			HNO_3, 4%		
	H_2O_2, 16%			H_2O_2, 16%		
Zn	HCl conc. 2 h	(1100)	27	HCl conc.	$V_{[1210]} > V_{[1100]} > V_{[0001]}$	26
	H_2O, 1 h	(0001)				

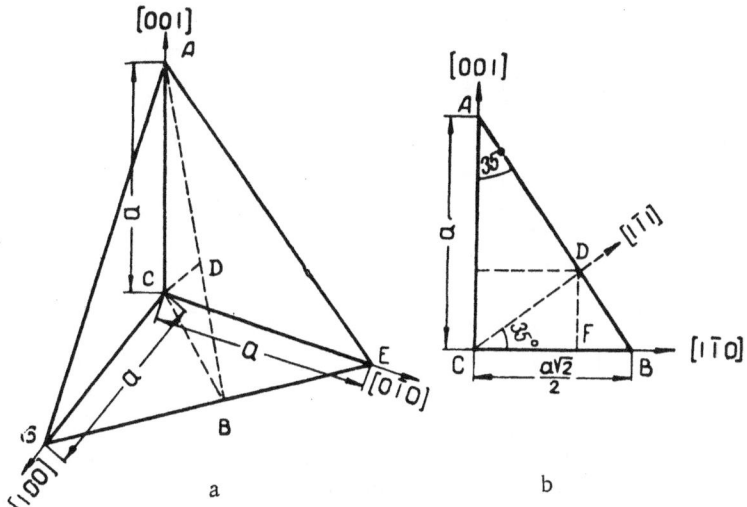

Fig. 1. Cubic etch figure on the octahedral plane (a), and its projection
on the rhombododecahedral plane (b).

the etch figures may most simply be considered for metal crystals with face-centered cubic lattices, since the dissolution of these crystals usually involves the revelation of the cubic or octahedral planes.

If the dissolution involves revelation of the cube faces, then the anisotropy of the dissolution rates for the most important directions in the crystal may easily be determined from Fig. 1. The point D corresponds to the center of development of an etch pit; DC is perpendicular to the plane of the octahedron. From the form of dissolution, and in particular from the triangle ABC, it is easy to determine the ratio of the dissolution rates for various crystal faces:

$$V_{[111]} : V_{[001]} = \frac{1}{\sin 35°16'} = 1.71,$$

$$V_{[111]} : V_{[110]} = \text{tg}\,35°16' = 1.23,$$

$$V_{[110]} : V_{[100]} = \text{ctg}\,35°16' = 1.41.$$

The order of decreasing dissolution rates will be as follows:

$$V_{[111]} > V_{[110]} > V_{[100]}.$$

If the dissolution of the crystal involves the revelation of the octahedral planes, then the ratio of the velocities changes. Figure 2 presents a scheme indicating the formation of such an etch pit, and triangle ABC, from which we can calculate the relative dissolution rates for the principal directions in the crystal (point B is the center of development of the etch figure):

$$V_{[110]} : V_{[100]} = 1.23,$$
$$V_{[100]} : V_{[110]} = 1.41,$$
$$V_{[100]} : V_{[111]} = 1.71.$$

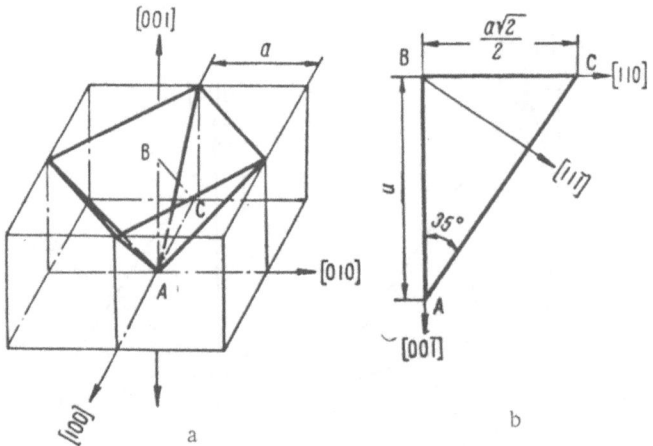

Fig. 2. Octahedral etch figure on the cube plane (a) and its
projection on the rhombododecahedral plane (b).

The order of decreasing dissolution rates is

$$V_{[100]} > V_{[110]} > V_{[111]}.$$

The calculations presented enable us to forsee the effect of the anisotropy in dissolution rates, if we know
the form of the etch figures. Data of Glauner and Glocker [25] on the variation in the dissolution rates of various
faces of a copper crystal in 11 etching solutions show that the relative dissolution rates fall neatly within the
calculated limits.

The calculated dissolution-rate ratios may be compared with the experimental results of etching aluminum crystals in 30% caustic soda solution and an $HF-HCl-HNO_3$ acid mixture [4].

In alkaline solution, the dissolution involves revelation of the (335) planes, the orientation of which does
not differ greatly from that of the octahedral plane. In the acid mixture, only the cubic planes are revealed.

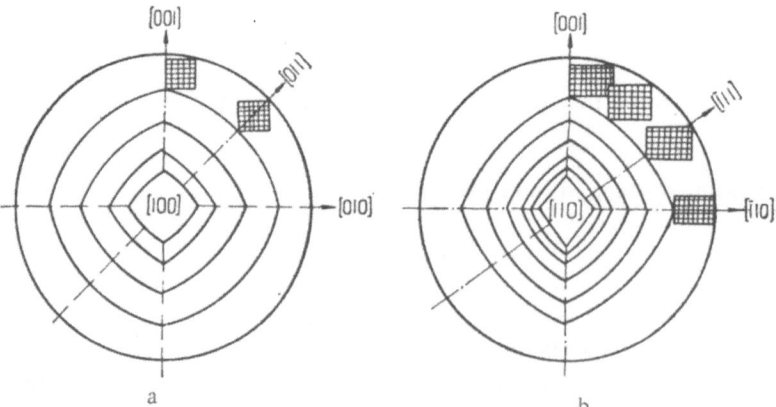

Fig. 3. Scheme of the dissolution of a single-crystal sphere under the
action of an etchant revealing the cubic planes: a) projection perpendicular to the emergence of pole [100]; b) projection perpendicular to
the emergence of pole [110].

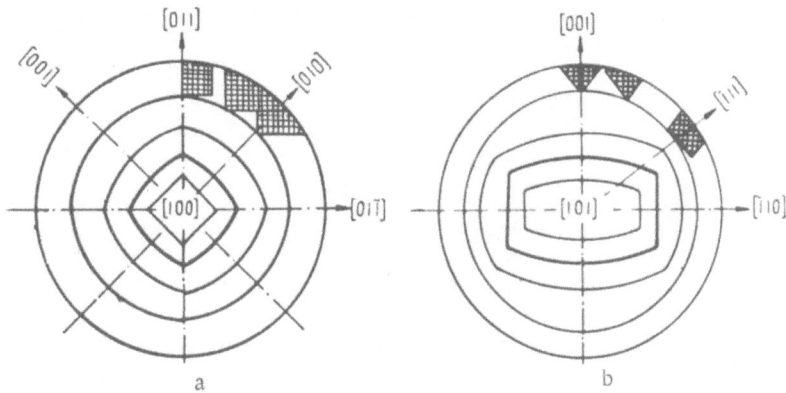

Fig. 4. Scheme of the dissolution of a single-crystal sphere under the action of an etchant revealing only the octahedral planes: a) projection perpendicular to the emergence of the [100] pole; b) projection perpendicular to the emergence of [110] pole.

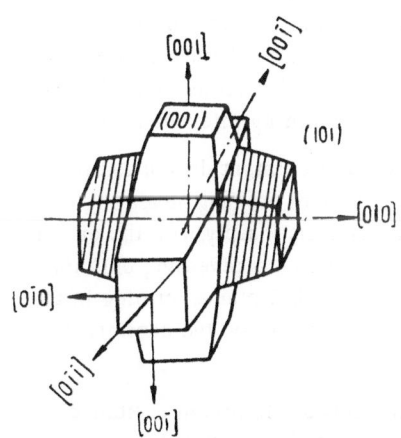

Fig. 5. Scheme of the final dissolution form when the sphere is dissolved so as to reveal the octahedral planes.

Comparison of the theoretical and experimental dissolution rates on etching aluminum crystals in alkali solution gives the following results: $V_{[110]}: V_{[111]}$ 1.23 and 1.24 (here and subsequently the first figure gives the ratio of the theoretical rates and the second that of the experimental), $V_{[100]}: V_{[110]}$ 1.41 and 1.12, $V_{[100]}: V_{[111]}$ 1.71 and 1.31.

On dissolution of aluminum in the acid mixture we have: $V_{[111]}: V_{[110]}$ 1.23 and 1.11; $V_{[110]}: V_{[100]}$ 1.41 and 1.15; $V_{[111]}: V_{[100]}$ 1.71 and 1.31.

As we see, agreement is quite satisfactory. The fact that the experimental anisotropy effect is smaller than the theoretical may be explained by the fact that the heavily etched surface contains a multitude of corners and edges which dissolve considerably faster than the face parts, and the possibility of some parts of the crystal being chipped off in the form of silt cannot be excluded.

The etch figures appearing in various forms on the crystal surface have a decisive effect on the form of dissolution of the crystal.

Figures 3a and b show projections of a crystal in the form of a sphere. For the case considered we suppose that the dissolution involves the revelation of the cube planes only. The projections of such etch figures on the cube and rhombododecahedral faces are shown by squares and rectangles respectively. The lines inside the circle correspond to the new surface of the crystal, which in the course of dissolution tends to assume the form of an octahedron. The almost-plane faces of octahedron appear after dissolution of $\frac{4}{5}$ of the crystal (reckoning by the diameter).

A completely different dissolution form is obtained from the crystalline sphere if etching involves the revelation of the octahedral faces. Figures 4a and b show the projection of the spherical crystal and etching figures on the rhombododecahedral and cube planes.

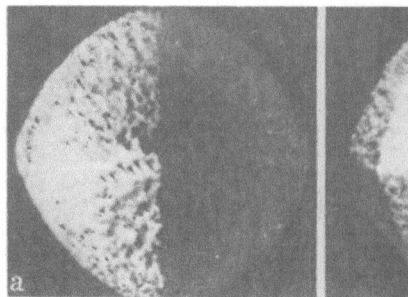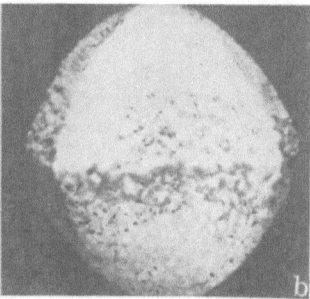

Fig. 6. Single-crystal sphere of aluminum dissolving in a mixture of hydrochloric, nitric, and hydrofluoric acids. Etching involves revelation of the cube planes: a) cube plane parallel to
the plane of the diagram ($\times 2$); b) rhomobododecahedral plane
parallel to the plane of the diagram ($\times 2$).

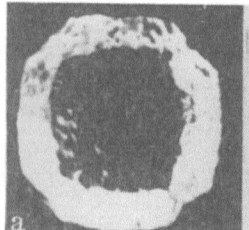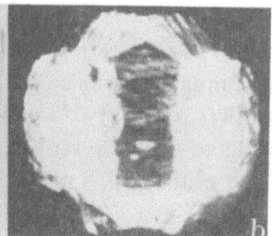

Fig. 7. Single-crystal sphere of aluminum dissolving
in 30% NaOH: a) cube plane parallel to the plane of
the diagram; b) rhomobododecahedral plane parallel
to the plane of the diagram.

The cube so arising does not constitute the
final dissolution form of the sphere, but rather some
intermediate state. On further dissolution of the crystal, its form approximates that shown in Fig. 5. The
formation of the cubes must take place after dissolution of $\frac{2}{3}$ of the sphere. The projections of the cube
in Fig. 4a and b are shown by thickened lines.

The absence of a constant dissolution form on
etching of the sphere with etchants revealing the
octahedral plane contrasts sharply with the dissolution of the crystal in etchants revealing only the cube
planes, since the octahedral planes forming in the
latter case do not vanish, but on the contrary perfect their form.

Comparison of these results with experimental data on the dissolution forms of aluminum crystals dissolving in 30% alkali solution and acid mixtures completely confirms the above laws. Figures 6 and 7 show photographs (taken from [4]) of the dissolution forms in the shape of octahedra and cubes. These dissolution forms
were obtained after etching away only $\frac{2}{3}$ of the original sphere. This fact in particular explains the rounded-
ness of the octahedral faces.

The dissolution rates of variously oriented crystal faces differ very little if dissolution involves the formation of etch figures bounded by planes with different indices. For example, in the dissolution of aluminum in
hydroiodic acid or in 5% HF, the surface relief of the metal consists of cubic and octahedral planes. In some
cases Engel [30] observed a similar picture for iron. In the dissolution of minerals also the phenomenon occurs
quite frequently.

Figure 8 shows an octahedron with conjugate cube faces and the projection of a spherical crystal subjected
to dissolution. Since the form of the etch figure to a first approximation differs little from a sphere, the dissolution of the sphere will take place in such a way that the newly forming surface will differ little from spherical. In practice such cases often occur with the dissolution of salts in unsaturated solutions. It is very likely
that this also explains the similar dissolution rates of different faces in rock-salt crystals.

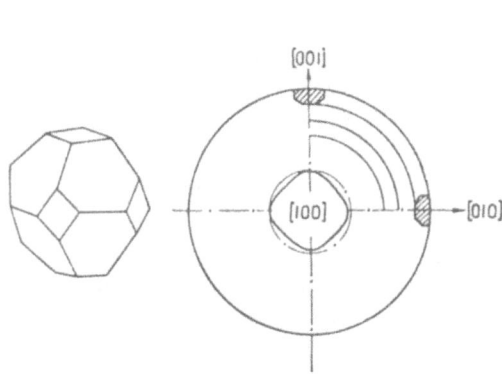

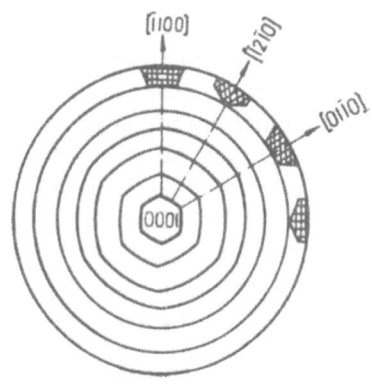

Fig. 8. Dissolution of a crystal sphere when the
etch figure has a form close to spherical.

Fig. 9. Dissolution of a single-crystal
cylinder of zinc.

Hence isotropic dissolution of a crystal is not caused by all etch figures, but only those similar in form to a sphere.

Data on the dissolution of a single crystal of zinc in hydrochloric acid [27] indicate that the dissolution of the side surface of a cylinder oriented with its base in the [0001] direction should result in the revelation of a prism of the first kind. Figure 9 shows the projection of such a cylinder and the corresponding projections of the etch figures. As seen in the figure, the final dissolution form of such a crystal of the hexagonal system will be a prism bounded by (1210) planes. Yamamoto [26] found, in the course of an experimental study of the final dissolution forms of crystals and the mathematical analysis of the results, that it was precisely these dissolution shapes of single-crystal zinc cylinders that developed on dissolution in hydrochloric acid of various concentrations.

The placing of the emphasis upon etch figures in the study of crystal-dissolution processes enables us to answer yet another important question in the theory of crystal dissolution, namely, the reason for the continuity in the dissolution rates on passing from face to face.

Kuznetsov, in the book, "Crystals and Crystallization," has stated in regard to this problem: "... in dissolution we must not speak of the dissolution of any face as a plane. Having become covered with etch figures, the face is in essence a whole set of faces with different crystallographic characters, and the dissolution rates of these are very different. This apparently also explains the continuity of dissolution rates on passing from one face to another" (p. 119).

It follows from this that the continuity of dissolution rates results from the continuity of the statistical dissolution rates of different faces of the etch figures on passing from one, originally taken for the etching of the crystal face, to the other. This explanation, however, is still far from clear. If we assume that the etching of the faces of a particular crystal results in etch figures bounded only by particular planes, for example the cubic or octahedral surfaces, then it would appear that the mean statistical dissolution rate should be the same for all initially taken faces. This is in fact not so. It is necessary to take into consideration the orientation of the etch figures on the crystal faces, and this differs for different faces. As the examples with the dissolution of spherical crystals show, this leads to a continuous variation of the dissolution rates of the faces with crystallographical direction.

It should be noted that, when we speak of the similarity or difference of growth and dissolution processes in crystals, then by "face" we understand a large, macroscopic section. This is valid in most cases of crystal growth, but is quite inapplicable to cases of dissolution when the crystal faces with definite indices are only present on the scale of etch figures. Hence, strictly speaking, the rate of the process changes discontinuously on passing from face to face for dissolution as well as for crystal growth.

Literature Cited

1. A. Verma, The Growth of Crystals [Russian translation] (IL, 1958).
2. A. Politycki and H. Fischer, Z. Elektrochem. 56(4):326 (1952); 57(6):393 (1953).
3. M. Allbutt, J. Inst. Metals 88(4):190 (1959).
4. H. Orem, J. Res. Nat. Bur. Stand. 58(3):157 (1957).
5. H. Kostron and E. Höffler, Z. Metallkunde 144(1):17 (1953).
6. Z. Nischiyama and R. Fujita, Mem. Inst. Sci., Industr. Res. Osaka Univ. 15:113 (1958).
7. B. Rivolta and L. Peraldo Bicelli, Metallurgia Ital. 50(11):495 (1958).
8. A. Gwathmay and A. Benton, J. Phys. Chem. 44(1):35 (1940).
9. R. Piontelli, G. Poli, and B. Rivolta, Atti Accad. Naz. Lincei Rend. Cl. Sci. Fiz. Mat. e Natur. 26(3):321 (1959).
10. R. Piontelli and G. Poli, Z. Elektrochem. 62(3):320 (1958).
11. Gilman and Johnston, Collection: "Elementary Processes of Crystal Growth" [Russian translation] (IL, 1959, p. 249).
12. L. Cavallaro and I. Bolognesi, P. Rev. Metallurgie, 52(9):706 (1955).
13. E. Biderman, Z. Metallkunde, 50(8):481 (1959).
14. I. M. Novosel'skii, Dissertation (Kazan, 1962); G. S. Vozdvizhenskii and I. M. Novosel'skii, Transactions of Interuniversity Conference on the Anodic Protection of Metals (Kazan, 1962).
15. N. P. Pavlov, Proceedings of the Novorossiisk Society of Natural Research 25(2):83 (1904).
16. G. S. Vozdvizhenskii, Trudy Kazansk. Khim. Tekhnol. Inst. (13):28 (1948); (14):26 (1949); Dokl. Akad. Nauk SSSR 59(9):1587 (1948). Transactions of the Second All-Union Conference on Theoretical and Applied Electrochemistry (Kiev, 1949), p. 80; Transactions of the Conference on Electrochemistry (Moscow, 1953), p. 410.
17. G. V. Akimov, Izv. Akad. Nauk SSSR, Otd. Khim. Nauk 5:469 (1951).
18. N. S. Akhmetov, Trudy Kazan'sk. Khim. Tekhnol. Inst. (19-20):261 (1954-55).
19. G. S. Vozdvizhenskii, V. A. Dmitriev, and E. V. Rzhevskaya, Zh. Fiz. Khim. 29(2):280 (1955).
20. B. N. Bushmanov, Master's Dissertation (Kazan, 1958).
21. H. Mahl and I. N. Stranski, Z. Phys. Chem. B51:319 (1942).
22. C. Walton, J. Trans. Electrochem. Soc. 85:239 (1944).
23. Kamiya, Ref. Zh. Khim. (1956), p. 4, Abstract No. 11,654.
24. H. Leidheiser and A. T. Gwathmay, Trans. Electrochem. Soc. 91:95 (1947).
25. R. Glauner and R. Glocker, Z. Krist. 80(5):577 (1931).
26. M. Yamamoto, J. Japan Inst. Metals 19(10):595 (1955).
27. M. Straumanis, Z. Krist. 75(5-6):430 (1930).
28. Yu. A. Skakov, Zavodskaya Laboratoriya 23(7):806 (1956).
29. G. Buckley, Growth of Crystals [Russian translation] (IL, 1954).
30. H. Engel, J. Arch. Eisenhüttenwessen 26(7):393 (1955).

DISSOLUTION STRUCTURES OF INDIVIDUAL FACES OF ALUMINUM SINGLE CRYSTALS IN A SOLUTION FOR CHEMICAL POLISHING

V. A. Dmitriev, E. V. Rzhevskaya, and V. A. Khristoforov

In the majority of cases, the dissolution of a metal in electrolytes is associated with the work of local microelements. The existence of such microelements is due to the electrochemical heterogeneity of the metal surface. As a result of this kind of dissolution, various etch figures appear on the metal surface, the external form of these depending on a number of interrelated factors.

On the one hand, the heterogeneity of a metal surface is due to the structure of the metal, the presence of various grains which differ in composition and orientation, intergrain boundaries, internal stresses, nonuniform distribution of impurities, various distortion of the crystal lattice of the metal, and to various other causes. On the other hand, the appearance of a heterogeneous nature of a metal surface depends to a great extent on the medium which the metal touches and with which it interacts [1].

One of the clearest examples of such dependence can be seen in both chemical etching and chemical polishing processes.

Both processes are electrochemical and involve the disruption of the crystal lattice of the metal at the metal—electrolyte interface, as a result of which a new surface structure develops. In chemical etching the structure of the metal is revealed, and here we may distinguish three types of localization of the process. These are macrolocalization, which consists of the sharp revelation of individual metal grains; microlocalization, in which the forms of the grains themselves are etched; and submicrolocalization, in which the fine structure of individual submicroscopic parts of the surface appear, these only being observable at electron-microscope magnifications.

In the study of the structure of a chemically polished surface, this subdivision of structure, however, presents considerable difficulty. At the present time sufficient systematic data are not available to enable us to judge the structure of a chemically polished surface. It is of theoretical and practical interest to obtain such data.

Some experimental data obtained in the study of the dissolution structures of polycrystalline and single-crystal aluminum are presented below.

In studying polycrystalline samples we used sheet aluminum of type A1M (Al—99.5%, Fe — 0.2, Si — 0.2, Cu — 0.015, and other impurities 0.085%) and AOO (Al — 99.7%, Fe — 0.16, Si — 0.1, Cu — 0.01, and other impurities 0.03%).

The single crystals were grown by crystallizing from a melt of AOO-type aluminum.

The microscopic examination was made with magnification of × 500, and the electron-microscopic, × 23,000. In the latter case we used lacquer replicas shadowed with chromium.

The sample surfaces were preliminarily treated by electropolishing in a bath of the following composition (wt. %): H_3PO_4 — 30, H_2SO_4 — 33, CrO_3 — 4, H_2O — 33. The current density was 50 A/dm^2, the duration of polishing 10 min, and the temperature 90 to 95°C.

This treatment of the aluminum led to the formation of a visually and microscopically smooth surface and produced no structural changes on the metal surface.

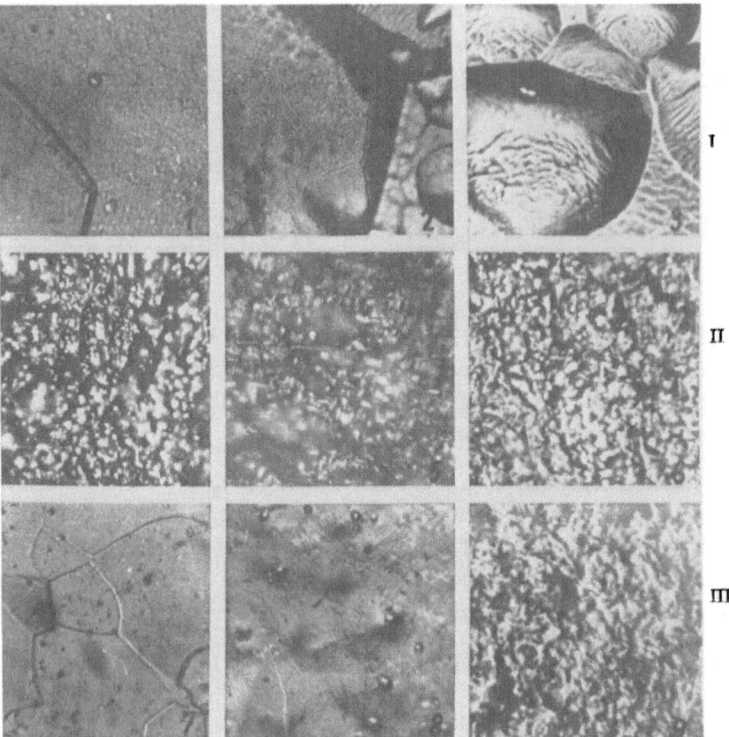

Fig. 1. Dissolution structure of polycrystalline aluminum in phosphoric acid. I) AOO aluminum; II) A1M (35% phosphoric acid, sp. gr. 1.21, temperature 98°C, 10 min (1, 4-20 min; 2, 5-120; 3, 6-180 min); III) A1M (7-87% phosphoric acid 50°C; 8-87% phosphoric acid, 98°C; 9-97% phosphoric acid, 98°C). ×500.

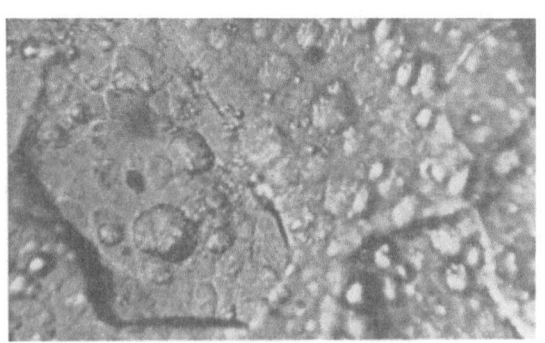

Fig. 2. Dissolution structure of polycrystalline aluminum in sulfuric acid. ×500.

For chemical polishing we used a solution composed of a mixture of concentrated acids: 78 ml H_3PO_4 (sp. gr. 1.71), 15 ml H_2SO_4 (sp. gr. 1.82), 7 ml HNO_3 (sp. gr. 1.49), and 1 g $Cu(NO_3)_2$, at a temperature of 98°C.

Dissolution structures were also studied in individual components of the chemical polishing bath (phosphoric and sulfuric acid).

Dissolution Structures of Aluminum in Phosphoric and Sulfuric Acids

The nature of the structure developing during the dissolution of polycrystalline aluminum in phosphoric acid depends on the duration of dissolution, the concentration of the acid, the purity of the metal, and the temperature (Fig. 1). Aluminum type A1M dissolves in 35% phosphoric acid with the formation of coarse relief, this etching character being preserved throughout the whole experiment (up to 3 h). Purer aluminum (AOO) behaves differently. Here the grain boundaries gradually develop over 1 h, the grains themselves dissolving uniformly, however. On longer dissolution, the intense development of the grain boundaries leads to the formation of microrelief consisting of rounded protuberances constituting the remains of crystallites. In more concentrated phosphoric acid (87%) at 50°C, microstructure appears, and at 98°C glossy etching takes place. On further raising of the concentration of the acid, the grains themselves are etched, without revealing their boundaries.

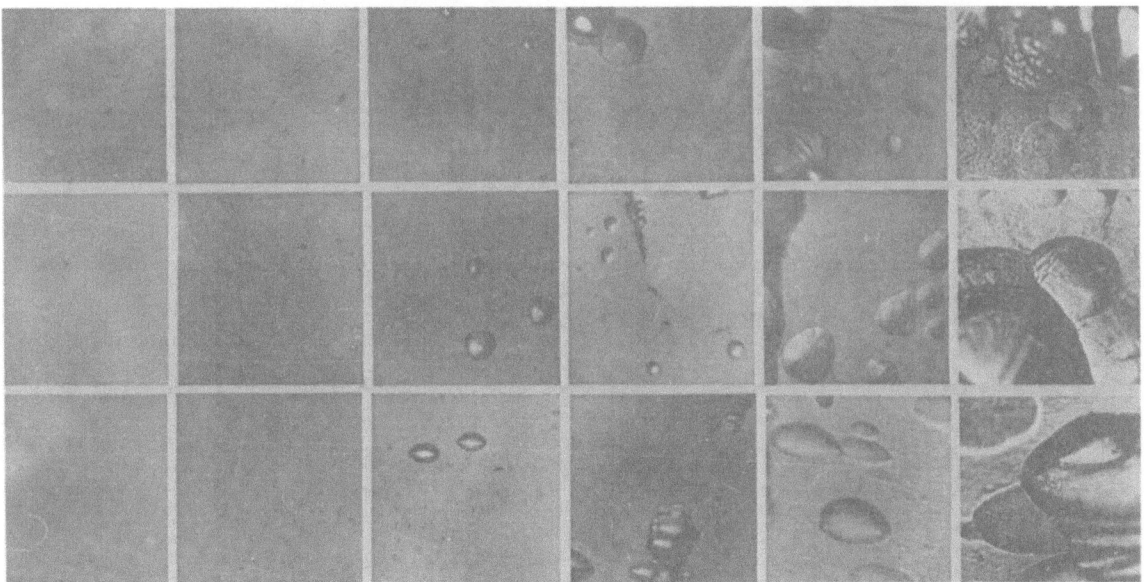

Fig. 3. Dissolution structures of individual faces of an aluminum single crystal in phosphoric acid. × 500.

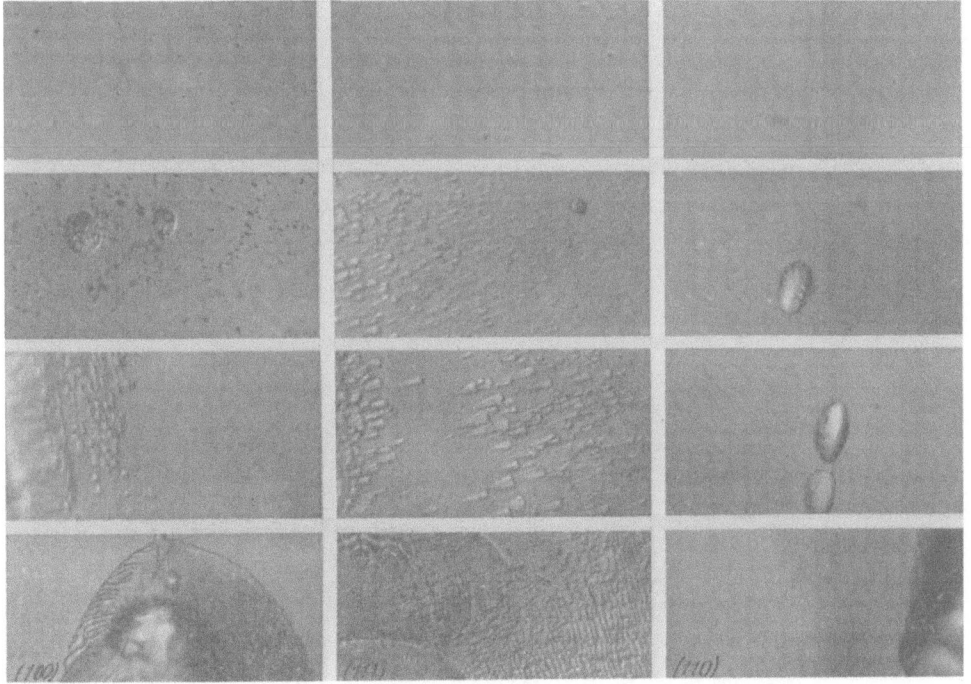

Fig. 4. Dissolution structures of individual faces of an aluminum single crystal in 94% sulfuric acid, 98°.

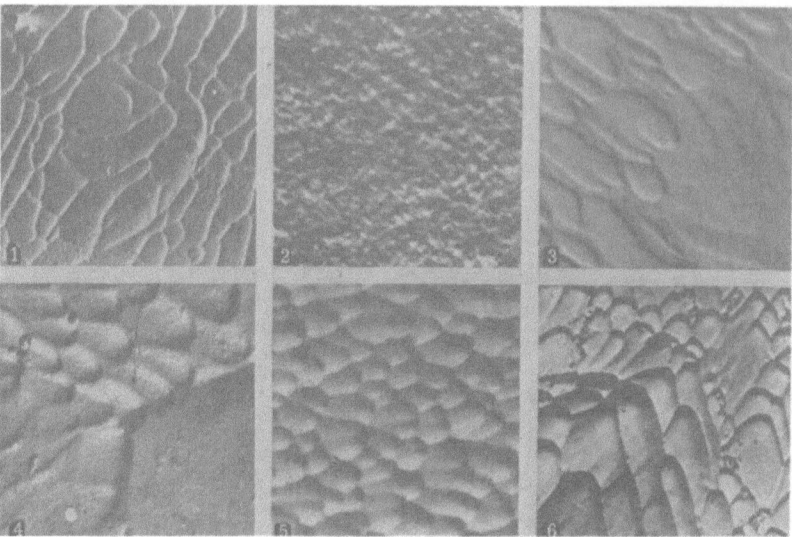

Fig. 5. Electron-microscope structure of polycrystalline aluminum after dissolution in sulfuric and phosphoric acids (for 10 min): 1) 94% sulfuric acid, 98°; 2) 79% phosphoric acid, 98°; 3) 79%, 50°; 4) 87%, 98°; 5) 87%, 50°; 6) 35%, 98°. × 14,720.

The dissolution of polycrystalline aluminum in 94% sulfuric acid at 98° is also accompanied by the revelation of microstructure (Fig. 2).

The dissolution of individual faces of an aluminum single crystal in 35% phosphoric acid at 98°C proceeds very uniformly for the first 30 min, and only on further extending the process does pitting appear on the surfaces and gradually increase.

The character of the etch figures forming is usually not connected with the crystallographic classification of the plane dissolving (Fig. 3). In 87% phosphoric acid the dissolution of the single crystal proceeds more intensely; as in 35% acid, regular dissolution figures are not formed.

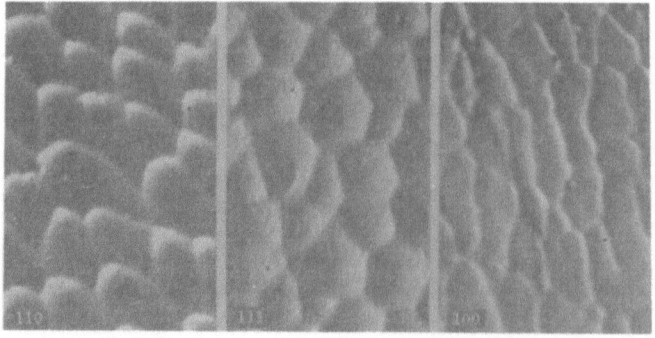

Fig. 6. Electron-microscopic structure of individual planes of an aluminum single crystal after dissolution in 35% phosphoric acid (98°, 20 min). × 23,000.

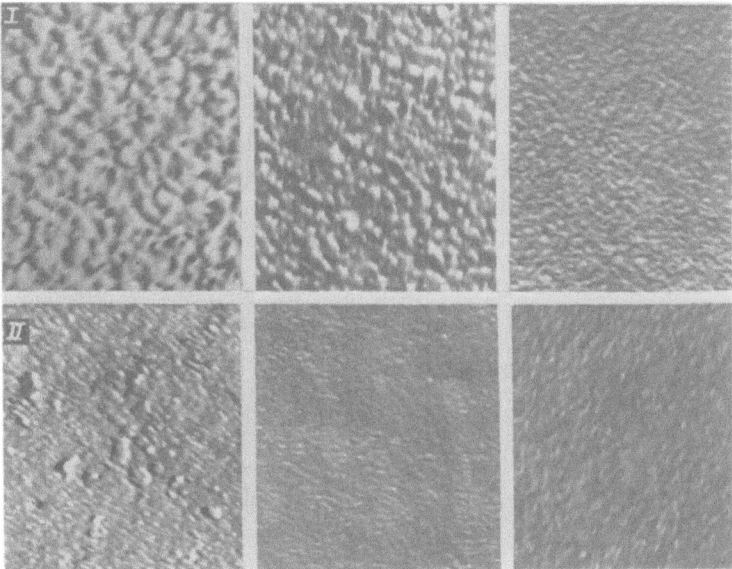

Fig. 7. Fine structure of an aluminum single crystal after chemical and electrolytic polishing: I) electropolishing; II) chemical polishing. x 23,000.

The single crystal dissolves rather differently in sulfuric acid (Fig. 4). The (110) plane dissolves more uniformly, and the coarsest structure develops on the (111). On continuation of the dissolution, the surfaces of all the planes are covered by coarse etch figures. From some of these the crystallographic classification of a given plane can be judged, but in the majority of cases the external appearance does not reflect the true crystal forms.

Study of the fine substructure of the surface of polycrystalline aluminum showed that on the submicroscale the dissolution of the metal had a local character (Fig. 5). In sulfuric acid a peculiar layer structure developed. The variety of structures developing in phosphoric acid depends on its concentration and temperature. The most typical structures forming on the (100), (111), and (110) planes appear in Fig. 6.

Thus on the submicroscale the dissolution of individual planes of a single crystal has a local character which to a certain extent is connected with the crystallographic orientation of the surfaces undergoing treatment. It should be noted, however, that quite often one also finds other types of structure similar to those which were observed on the surface of the polycrystalline metal. Hence the character of the etch developing figures depends not only on the crystallographic orientation, but frequently to a considerably greater extent on other factors. Chief among these are apparently concentration variations at the metal—solution boundary and passivating layers on the metal surface which can arise during the dissolution process. This can be seen especially clearly in the chemical polishing of aluminum.

Dissolution Structures in the Chemical Polishing Bath

The character of the developing micro- and submicrostructures does not change substantially even in the case when the metal dissolves in a mixture of phosphoric and sulfuric acids in the proportions which we used for the chemical polishing bath. A very different picture arises, however, when 6 to 8% nitric acid is added to this mixture.

With the dissolving of polycrystalline aluminum in such a solution, the surface takes on a glossy appearance. Microscopically, however, in contrast to the electropolished surface, traces of intergrain boundaries ap-

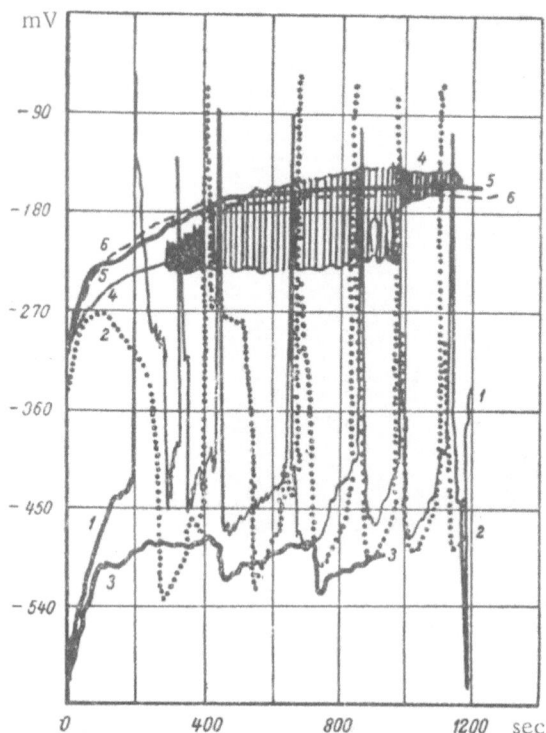

Fig. 8. Effect of nitric acid on the aluminum dissolution potential in a chemical polishing bath; nitric
acid content (vol = 1.51): 1) 1 vol. %; 2) 2 vol. %; 3)
0 vol. %; 4) 4 vol. %; 5) 6 vol. %; 6) 10 vol. %.

pear on it; the grains themselves, however, in all cases dissolve and form a homogeneous structure, which is not associated with the presence of differently oriented grains. On dissolution of a single-crystal sample, all three of the planes studied (cube, ocathedral, and rhombododecahedral) at microscopic magnifications show a smooth surface without traces of any regular etch figures.

Electron-microscopic study of such samples, however, permits the observation of fine surface structure. Identical substructure forms on all three planes after electrolytic polishing, constituting chaotically disposed relief.

After treatment of such surfaces in the chemical polishing bath, still finer substructure develops; just as in electropolishing, this comprises chaotic submicroscopic unevenness (Fig. 7).

What causes the removal of structural etching in the metal?

As mentioned earlier, such causes may include concentration changes at the metal—solution boundary and passivating layers at the surface of the metal being dissolved. The two factors act simultaneously and cannot be considered in isolation.

Concentration changes at the metal—solution boundary, even on dissolution in phosphoric and sulfuric acids, in some cases produce lustrous surfaces with the simultaneous revelation of grain boundaries and the block structure of the single-crystal faces.

For the complete suppression of structural etching, additional factors are needed. Passivating layers constitute such factors. These make the surface electrochemically more homogeneous and at the same time do not hinder the process of dissolution of the metal. The existence of this incomplete passivation of aluminum in chemical polishing solutions may be judged from the nature of the changes in electrode potential [2].

If the concentration of nitric acid in the chemical polishing bath does not exceed 4%, there occur periodic changes of potential, caused by the periodic formation and rupture of passivating layers (Fig. 8).

With an increase in the nitric acid content to 6%, the periodic phenomena cease, since the amount of passivator is sufficient for the formation of stable passivating layers.

Further increase in the content of the oxidizing agent has little effect on the curve of the potential time.

An analogous phenomenon occurs during prolonged use of the chemical polishing solution. In the fresh solution, the aluminum potential lies between −270 and −280 mV with respect to the saturated calomel electrode.

With an increase in the duration of operation, the solution loses nitric acid, and when the concentration has dropped to 4%, periodic phenomena (oscillations) begin to occur; these cease after the nitric acid content has fallen to 0.8 or 0.5%. The aluminum potential meanwhile becomes more negative and lies between −355 and −360 mV. The surface structure changes in keeping with the changes in potential.

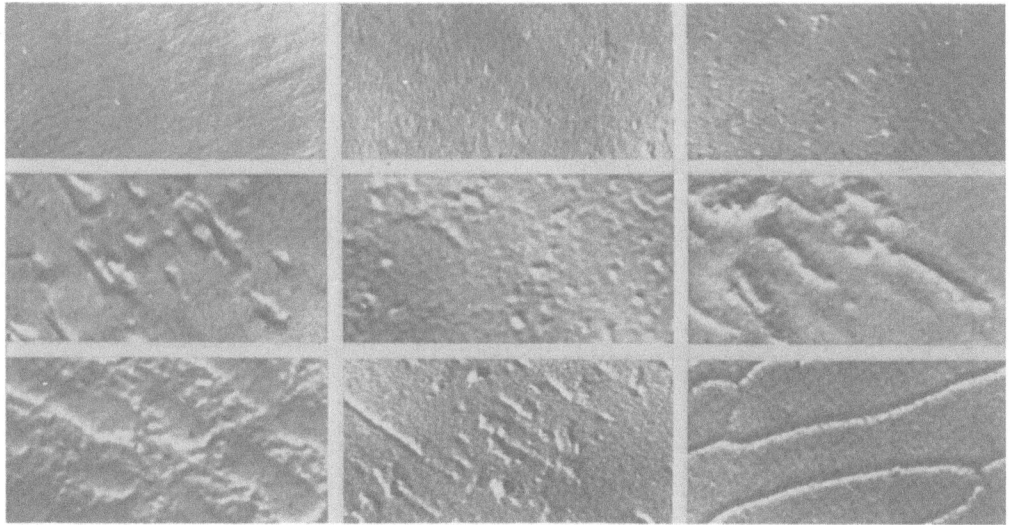

Fig. 9. Electron-microscope structure of an aluminum single crystal after chemical polishing:
I) before oscillations begin; II) during oscillations; III) after oscillations have ceased.

Macroscopic examination reveals the following: Before oscillations begin, a smooth, lustrous surface is formed; during the oscillations, small circular pits arise on all the single-crystal surfaces. If dissolution proceeds in a solution in which oscillations have already ceased, then coarse pits appear on the surface of the treated metal.

Microscopic examination of the surface of such samples reveals no new structural details.

Electron-microscope examination of a surface so treated shows that, whereas up to the onset of oscillations a fine substructure with a uniform distribution of unevenness is formed, during the oscillations local dissolution of the metal commences (Fig. 9). After the oscillations have ceased, the local character of the dissolution of individual single-crystal faces is still further strengthened.

The results obtained convincingly indicate that the presence of passivating layers of the surface of the metal being polished is an unfailing condition for the suppression of structural etching. When the concentration of the oxidizing agent is insufficient, the uniformity of passivation is disrupted, and electrochemical heterogeneity of the metal surface begins to appear.

These results clarify our ideas on the mechanism of the chemical polishing of aluminum.

Among the hypotheses designed to explain this process, the most widespread is that according to which one of the main conditions for producing a polished surface is the intense operation of local microelements on the surface of the dissolving metal [3]. Our data on the dissolution of structures of individual faces of an aluminum single crystal, however, give grounds for arriving at the opposite conclusion.

In fact, during chemical polishing, structural etching is suppressed, and the decrystallization of the metal under these conditions is characterized by an almost complete cessation of the operation of local microelements which also leads to structureless etching of the metal surface.

Literature Cited

1. A. I. Golubev, Corrosion Processes in Real Microelements (Oborongiz, 1953).
2. V. A. Dmitriev, Zh. Fiz. Khim. 36(6):1375 (1962).
3. H. Fischer and Koch, Z. Metall. 6(17/18):305 (1952).

ETCH SPIRALS ON SINGLE CRYSTALS OF STEEL

L. I. Lysak and B. I. Nikolin

In studying the effect of the conditions of growth for single crystals on the morphology and crystallography of the martensite transformations in steel, it was noted that, during electropolishing, spirals analogous to growth spirals developed either singly or in groups on the sample surface. In order to determine the nature of these spirals it was of interest to study the conditions on their formation and the surface relief.

Growth spirals have been observed both on natural and artifically grown crystals. The correct explanation of this phenomenon was only given in 1949 by Frank and Read; it was based upon the dislocation mechanism of crystal growth. According to their conception, growth spirals are formed during crystal growth as a result of the twisting of a rising step, and may be observed on a crystal surface directly under the microscope either during growth or after growth has ceased. In this way growth spirals were observed on many crystals of organic materials: copper, gold, silver, etc. In the majority of cases the growth spirals were observed on crystals grown from solution and obtained by electrolytic deposition or vacuum evaporation. The step height of the growth spirals is normally 10 to 2000 A. They cannot be observed directly on crystals grown from the melt, except for the case of transparent organic substances.

In such cases, indirect methods of etching, evaporation, or oxidation are used [1-3]. The step height of the spirals is 100 to 2000 A. There are not sufficient grounds, however, to suppose that the appearance of spirals on etching, evaporation, or oxidation constitutes direct proof of spiral growth of crystals from the melt.

We have observed spirals formed on electropolishing single-crystal and polycrystalline steel of the G12 types and the alloy G20. The single crystals and coarse crystalline samples were grown in a Tamman furnace in a neutral medium (argon), and also in an induction high-frequency furnace. After crystallization the alloy was homogenized at $t = 1200 \pm 50°C$ for 2 h and then water-quenched or air-cooled, so that at room temperature the austenite structure was obtained. Samples in the form of plates 1 to 2 mm thick were cut mechanically and etched electrolytically to remove the deformed layer. On further electropolishing, spirals appeared on the sample surface for certain electropolishing conditions only: $U = 50$ to 50 V; $I = 30$ A/dm^2; duration more than 20 sec.

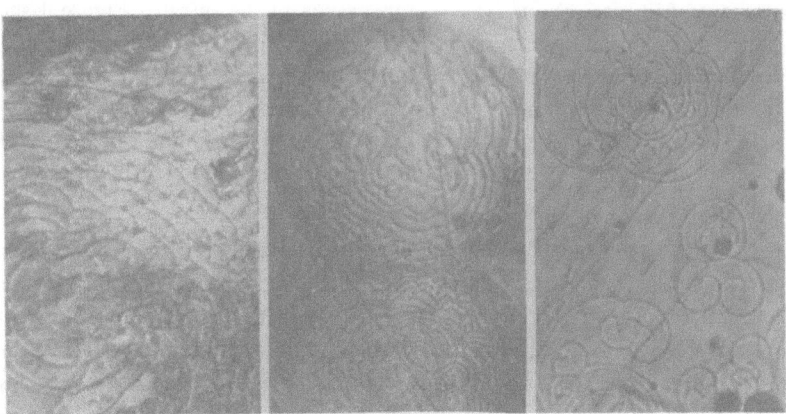

Fig. 1. Microstructure of various parts of a steel Fe−Mn−C single crystal grown from the melt after electropolishing. ×300.

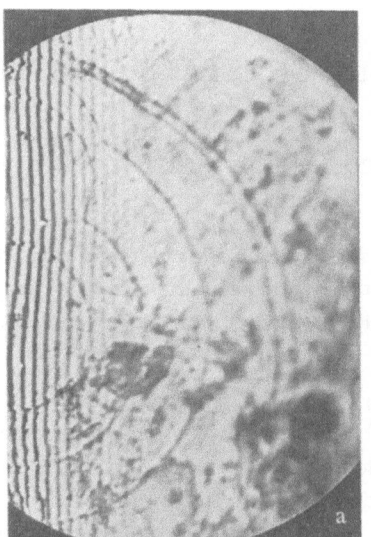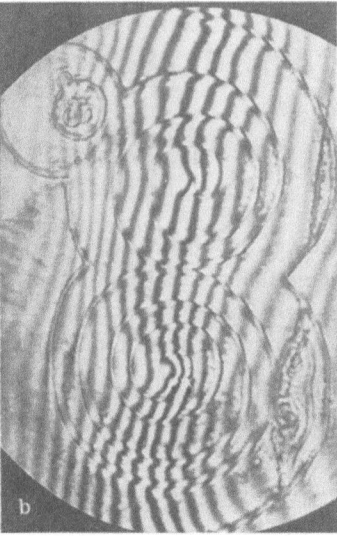

Fig. 2. Surface profile of Fe−Mn−C steel after electropolishing, as observed in the MIN-4 interferometer (× 500). a) Without a depression in the center of the spiral; b) with such a depression.

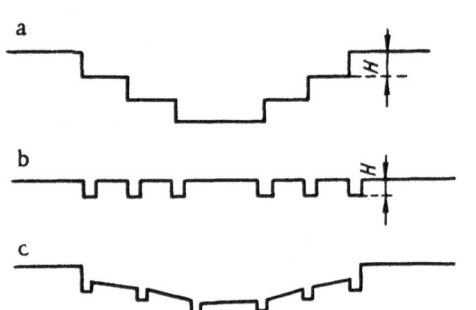

Fig. 3. Scheme of surface profile after electropolishing: a) in the presence of growth spirals; b) observed in Fig. 2a; c) observed in Fig. 2b.

The electropolishing of all the G12 steel and G20 alloy samples was effected in an electrolyte of the following composition: $\frac{3}{4}$ C_2H_5OH and $\frac{1}{4}$ $HClO_4$. It transpired that spirals are only formed for a certain current density, somewhat smaller than the current density for which electropolishing takes place. For other current densities either etching or polishing of the surface took place without the formation of spirals. It was established that the spirals were always formed on the surface indepently of the form and thickness of the sample, as the reduction in thickness on etching had no effect on the appearance of the spirals. Figure 1 shows photographs of spirals which in outward form are very similar to the growth spirals observed by Verma on silicon carbide [4] and on the dislocation sources of Frank and Read. We studied the profile of a surface on which spirals were visible in the Linnik MIN-4 interferometer. Figure 2a shows a profile record on which the profile of the spiral constitutes a groove 1000 to 2000 A deep and is shown schematically in Fig. 3b.

Figure 2b shows the profile of the surface of another spiral in which, apart from the observed groove, there is a considerable overall depression from the periphery to the center of the spiral. This profile is shown schematically in Fig. 3c. It is possible that the depression is connected with local etching of the sample. Figure 4 shows spirals formed on the surface of a steel single crystal obtained by recrystallization, and Fig. 5 shows spirals on the surface of a sample subjected to forging at t = 1100°C with 200 to 300% working.

The general form of these spirals does not differ from those forming on single crystals grown from the melt. It follows from this that the observed etch spirals do not depend on the structure of the crystal on which they are formed, and hence they neither constitute growth spirals nor are connected with dislocations of Frank−Read sources, as suggested in [5].

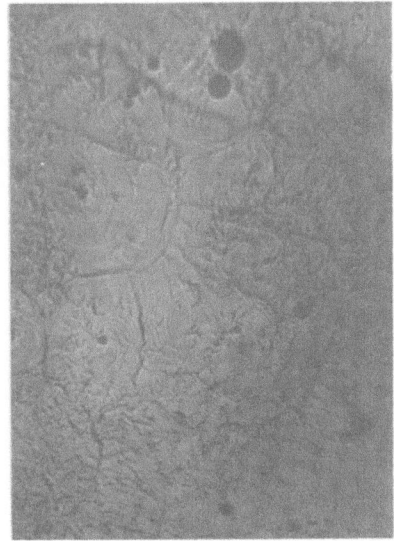

Fig. 4. Microstructure of an Fe−Mn−C single crystal obtained by recrystallization after electropolishing. ×300.

Fig. 5. Microstructure of Fe−Mn−C steel after forging and electropolishing. ×300.

This conclusion derives from a study of the profile of surfaces on which the spirals are formed. From Figs. 2a and b and 3 we see that the spiral constitutes a groove instead of a spiral step, as would be expected in the presence of a growth spiral (Fig. 3a).

Such spirals are also observed on electropolishing aluminum [6]. The authors of [6] suggest that they are formed as a result of the deposition of oxides on the sample surface and are connected with neither dislocations of Frank−Read sources nor growth steps.

Literature Cited

1. G. Bassi and J. T. Fourie, Acta Metallurgica 5(11):676 (1957).
2. A. J. Forty, Phil. Mag. 43:481 (1952).
3. F. W. Young, J. Appl. Phys. 27(5):554 (1956).
4. T. Verma, Dislocations and Crystal Growth [Russian translation] (Moscow, 1958).
5. S. Feliu and G. Castro, Acta Metallurgica 10(5):554 (1962).
6. J. H. Marchin and G. Wyom, Acta Metallurgica 10:10 (1962).

EFFECT OF THE pH OF THE SOLUTION ON THE FORM OF AMMONIUM DIHYDROPHOSPATE CRYSTALS

I. M. Byteva

It is known that the shape of ammonium dihydrophosphate crystals reacts sharply to the acidity of the solution [1-3]. For a slight supersaturation, longer crystals grow from a solution with pH = 3.5 and below than from solutions with higher pH (Fig. 1). In this paper we make an attempt to explain the effect of the pH of the solution of the shape of $NH_4H_2PO_4$ crystals, based on their structure and on processes taking place in solution.

Crystals of $NH_4H_2PO_4$ have a body-centered tetragonal lattice. The space group of the crystals is $D_{2d}^{12} = I_4^-2d$, c = 7.516 A, a = 7.479 A [4]. The PO_4^{3-} groups are interconnected by hydrogen bonds. Considering the growth of ammonium dihydrophosphate crystals, we shall suppose that the structural units in the crystallization process are NH_4^+ and $H_2PO_4^-$ ions. We shall ascribe the phosphorus-atom sites in the lattice to the latter. We shall regard the ions as point charges. The space lattice constructed on such units is shown in Fig. 2.

On comparing the external shape and internal structure of the crystal, we see that the plane (011) networks are constructed of monotypic NH_4^+ or $H_2PO_4^-$ ions. The prism-face networks (010) consist of alternating regions of positive and negative ions, c/2 in breadth, parallel to the [100] direction.

Using the method proposed by Hartman and Perdok [5], we can calculate the crystal habit theoretically. The calculation amounts to the fact that the growth rate is proportional to the energy required to add a molecule to a given face. This energy was defined by Hartman and Perdok as the energy of the bond freed when the structural unit is linked to the surface of the growing face. For our case we must determine E_{011}/E_{010}. In order to calculate energy E_{010}, let us split the crystal structure into layers parallel to the (010) and calculate the electrostatic potential U_1 at the site of the NH_4^+ ion created by the nearest (010) layer. For this we sum the potentials created by all the [001] rows of this layer. The total potential we find by summing the potentials of all the layers parallel to the (010).

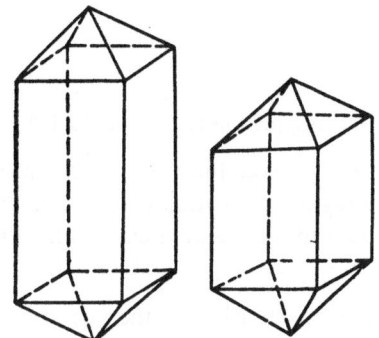

Fig. 1. Form of ammonium dihydrophosphate crystals grown from solutions with various pH.

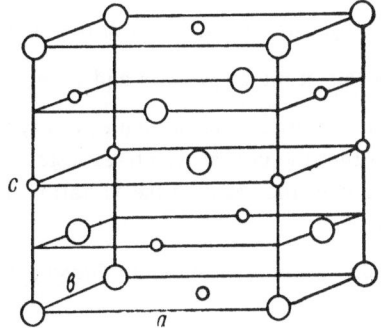

Fig. 2. Lattice of $NH_4H_2PO_4$.

The potential of a discrete row of charges was calculated by Madelung [6]:

$$U = \frac{4q}{p} \sum_{l=1}^{\infty} k_0 \left(\frac{2\pi \, lr}{p} \right) \cos \frac{2\pi \, lz}{p} + \frac{2q}{p} \ln \frac{2p}{r} + c,$$

where q is the charge of the ion, p the period of the row, r and z the cylindrical coordinates of the ion, and k_0 a Hankel function.

At the site of the positive ion, the potential from the nearest layer is

$$U_1 = - \frac{4q}{c} \, 0.0165 \text{ A}^{-1},$$

from the second

$$U_2 = \frac{4q}{c} \, 0.0008 \text{ A}^{-1},$$

and the third

$$U_3 = - \frac{4q}{c} \, 0.00001 \text{ A}^{-1}.$$

The total potential is

$$U_{010} = - \frac{4q}{c} \, 0.026 \text{ A}^{-1}.$$

In our case the charges of the ions are equal to the unit charge, and the energy of application of a molecule to the (010) face is

$$E = 2 \, eU_{010} = 0.027 \, e^2 \text{ A}^{-1}.$$

The correction for the nonspherical form of the H_2PO_4 ion [5] equals $0.064 \, e^2 \text{ A}^{-1}$, so that $E_{010} = 0.091 \, e^2 \text{ A}^{-1}$. For E_{011} we obtain $0.178 \, e^2 \text{ A}^{-1}$ per molecule.

The habit of $NH_4H_2PO_4$ crystals is determined by the ratio

$$\frac{h_{011}}{h_{010}} = \frac{V_{011}}{V_{010}} = \frac{E_{011}}{E_{010}} = \frac{0.178}{0.091} = 1.95.$$

where h_{011} and h_{010} are the distances of the faces from the growth center.

For crystals grown at pH = 3.5 and lower, $h_{011}/h_{010} = 2.1$; with increasing pH the ratio falls to 1.0.

In order to compare the structures of crystals grown at different pH, we examined their Debye photographs. These showed that the crystal structures were identical. Hence the energy of application will also be the same. In order to discover the cause of the variation in outward form, let us turn to the medium out of which the crystal grows.

The dissociation of ammonium dihydrogen phosphate in water may be described by the equations

$$NH_4H_2PO_4 \rightleftarrows NH_4^+ + H_2PO_4^-,$$

$$H_2PO_4^- \rightleftarrows H^+ + HPO_4^{--},$$

$$HPO_4^{--} \rightleftarrows H^+ + PO_4^{3-},$$

$$K_2 = 6.23 \cdot 10^{-8},$$
$$K_3 = 2.2 \cdot 10^{-13},$$

(1)

where K_2 and K_3 are the second and third dissociation constants of orthophosphoric acid. The $NH_4H_2PO_4$ is a strong electrolyte, and the dissociation indicated in Eq. (1) proceeds almost to completion. We shall subsequently neglect the dissociation of the HPO_4^- ion in view of the smallness of K_3.

The NH_4^+ and $H_2PO_4^-$ ions form hydrates of different stability with water. The heat of hydration of the NH_4^+ ion is 79 kcal/mole. The heat of hydration of the $H_2PO_4^-$ ion may be calculated. For this we determine the lattice energy from Kapustin's formula

$$E = 287.2 \ \Sigma n \ \frac{z_1 z_2}{r_1 + r_2} \left(1 - \frac{0.345}{r_1 + r_2} \right),$$

where Σn is the number of ions in the molecule, z_1 and z_2 are the valences of the ions, r_1 and r_2 are the radii of the ions for the NaCl-type lattice with coordination number 6. For $NH_4H_2PO_4$, $r_1 + r_2 = 2.7$ A.

$$E = 182 \ \text{kcal/mole}.$$

The heat of solution of $NH_4H_2PO_4$, ΔH, is -4.3 kcal/mole. From the law of the conservation of energy $E = H_1 + H_2 + \Delta H$, we find the heat of hydration of the $H_2PO_4^-$ ion:

$$H_2 = 107 \ \text{kcal/mole}.$$

Stroitelev showed [7] that the strength of the hydrates considerably influenced the habit of the crystals. The $H_2PO_4^-$ ions are more strongly bound to the solution than the NH_4^+ ions, and it is they which give a growth rate smaller than theoretical for the (011) face, since considerable work is required to join them. On the (010) face the hydration of $H_2PO_4^-$ has less effect, since the OH is joined to the network together with the NH_4^+ ion, which enters the lattice relatively easily. If we consider only the energy of hydration, then the (011)-face growth rate is determined by the quantity 79 kcal/mole, and the (010)-face by 107 kcal/mole; the ratio of the (011) and (010) face growth rates will equal

$$\frac{V_{011}}{V_{010}} = 1.4.$$

Since in an actual case both factors act, the velocity ratio becomes 1.9.

Let us consider what happens on adding NH_4OH to the solution. As a result of the dissociation $NH_4OH \rightleftharpoons NH_4^+ + OH^-$, additional OH^- ions are formed, and these, interacting with H^+, increase the pH of the solution. The $H_2PO_4^-$ and HPO_4^{--} ions, uniting with NH_4^+, give $NH_4HPO_4^-$. The equilibrium described by Eq. (1) shifts toward the right. A first result of this is that the solubility of the substance should increase, and this is indeed observed in practice. Figure 3 shows solubility curves of the substance for various pH of the solution.

The ions of the crystal which produce the growth rate of the pyramid faces become bound into relatively stable complexes on addition of alkali to the solution. The free ions become fewer than in the pure solution, and in order to extract them from the $NH_4HPO_4^-$ complexes extra work is needed. This reduces the number of particles which may enter the crystal, and V_{011} falls.

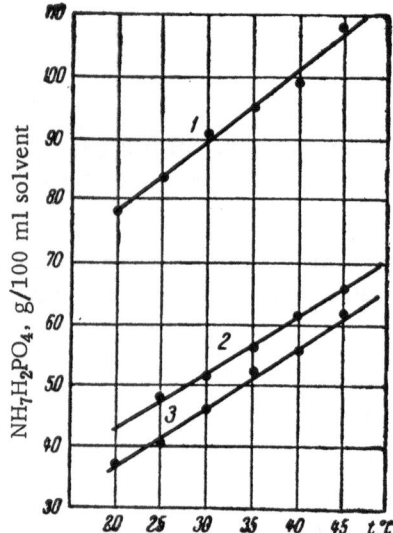

Fig. 3. Solubility of $NH_4H_2PO_4$: 1)
pH = 6; 2) 1.8; 3) 3.5.

On adding H_3PO_4 to the solution, the solubility of ammonium dihydrophosphate also increases. This happens because of the formation of the compound $NH_4H_2PO_4 \cdot H_3PO_4 \cdot H_2O$ at low pH [8], which binds the $H_2PO_4^-$ and NH_4^+ ions. However, the complexing in this case is far less, and its effect on crystal growth is insignificant.

Changes in the pH of the medium affect a number of other crystals in the same way: no new faces appear, but the relative sizes of existing faces change. Thus crystals of $Li_2SO_4 \cdot H_2O$ grow as planes on adding sulfuric acid to the solution, KH_2PO_4 crystals react to pH variations in the same way as $NH_4H_2PO_4$, triglycine sulfate crystals become longer, and so forth. There are grounds for thinking that in these cases the changes in growth form may be explained largely by complexing in solution.

Literature Cited

1.   H. Buckley, Growth of Crystals [Russian translation] (IL, 1954).
2.   G. Jaffe and B. Kjellgren, Collection: "Growth of Crystals" [Russian translation] (IL, 1950).
3.   I. M. Byteva, Collection: "Growth of Crystals" [in Russian], Vol. 3, p. 296 (1961); English translation: New York, Consultants Bureau, 1961.
4.   I. West, Z. Kristallog. 74(3-4):306 (1930).
5.   P. Hartman and W. G. Perdok, Acta Cryst. 8:49 (1955); 9:569 (1956); 9:721 (1956).
6.   E. Madelung, Phys. Zeitschr. 19:524 (1918).
7.   S. A. Stroitelev, Byull. Nauchn.-Tekhn. Inf., No. 3 (TsNIIOlovo, Novosibirsk, 1961).
8.   B. A. Muromtsev and L. A. Nazarov, Izv. Akad. Nauk SSSR, Seriya Khim., Vol. 1 (1938).

GROWING ALKALI-HALIDE SINGLE CRYSTALS
FROM THE MELT BY DIRECTIONAL HEAT EXTRACTION

A. E. Malikov

Until recently, large alkali-halide single crystals were used in infrared spectroscopy and scintillation techniques. The main requirements for such crystals were macroscopic and optical homogeneity and a given volume distribution of activator (for ensuring the necessary technical parameters, transmission bands, scintillation time, etc.). A new sphere of application for the single crystals (sound conductors for ultrasonic delay lines [1]) makes increased demands on the mechanical homogeneity of the material, since sound conductors require maximum transmission of an assigned frequency band with minimum scattering losses, absorption of ultrasound, and distortion of the ultrasonic pulse shape. For acoustically homogeneous single crystals, the propagation characteristics of plane acoustic waves in a given direction must not vary from point to point. A problem directly arising in this connection lies in the perfecting of the internal structure of the single crystals by improving the methods of crystallization and annealing.

The solution of this problem implies the development of methods for purification of the original raw material, by producing a perfect seed and growing it, and selecting the best conditions for growing and annealing the single crystals, keeping due control of their quality.

This paper describes our proposed system for the production of single crystals from the melt by the directional heat-extraction method (see the figure).

The furnace consists of a ceramic cylinder 4, on the outside of which is wound an electric heater 5 made of ÉI-626 alloy. The coil is wound nonuniformly in such a way as to raise the temperature by 5 to 7°C for each centimeter of height.

The most heated zone lies at 160 mm above the lower end of the ceramic cylinder. The turns are distributed more sparsely at the top, thus ensuring a sharp temperature drop in the upper part of the furnace.

The heater winding is covered with a paste consisting of a mixture of Al_2O_3, clay, and chamotte particles in the ratio 5:2:3. This layer prevents the helix from slipping when heated. The remaining volume of the furnace is filled with heat-insulation layers 11 and 12.

In the lower part of the working chamber, through an opening in the base, is set a condenser bearing a stainless steel disk. The crucible 1 is placed with its cone in a depression in the disk, and is covered from above with a lid. In the lid is set a porcelain tube. In the upper part of the furnace are a number of gaskets 7, 8, and 9.

On to gasket 9 is poured a layer of activated charcoal 10, intended to absorb chlorine and bromine ions evolving from the mixture on melting. This plugging of the upper part of the furnace makes it possible to work with NaCl, KCl, and KBr without a fume cupboard.

The growth of the single crystals is effected in a stationary furnace with the crucible in a stationary state.

The porcelain vessel (crucible) is charged with the salt and placed in the furnace. The furnace is heated until the salt is completely melted. After this, the melt is superheated to 150 or 200°C and held there for 1 h (temperature control effected by the readings of a thermocouple set in the crucible 2 cm from the surface of the melt). Air is then blown through the inner tube of the condenser. After 30 min the air is replaced by water. The initial water flow is 12 liter/h; it then gradually increases and at the end of the experiment reaches 180 liter/h.

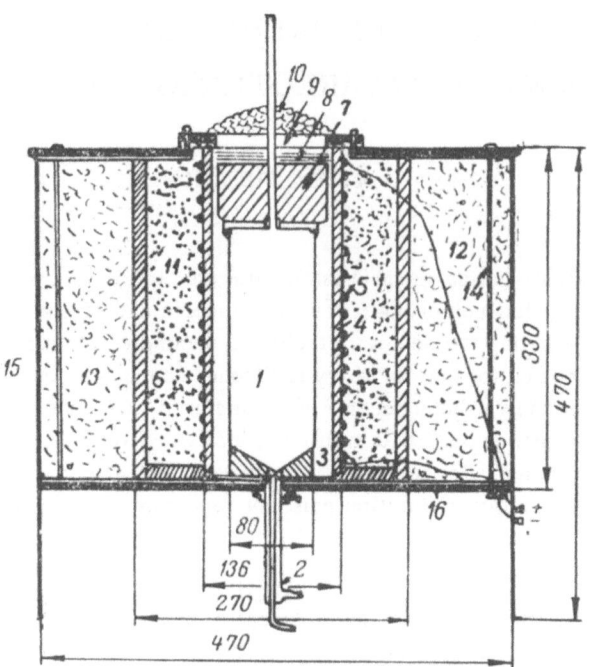

Arrangement of the furnace for growing single crystals by
the directional heat-extraction method: 1) porcelain crucible; 2) heat-extraction system (cooler); 3) metal disk; 4)
ceramic tube; 5) heater winding; 6) ceramic clyinder; 7)
gasket of chamotte bricks; 8, 9) nickel gasket; 10) activated
charcoal; 11) chamotte pieces; 12, 13) asbestos pieces; 14)
tightening screws; 15) sheath; 16) iron base.

After the melt has crystallized completely, the furnace is gradually cooled to room temperature.

Thus, while the current required by the furnace heater is kept constant, the crystal grows as a result of
directional heat extraction.

The spontaneous generation of a single center in the conical part of the crucible is explained by the con-
structionally created possibility of extracting heat in the first instance from the sharp conical end of the crucible. Further growth of the single crystal is effected by gradually raising the heat extraction from the lower
part of the crystal.

By this method we obtained single crystals 100 mm in diameter and 130 to 140 mm long. The growth
rate of these crystals reached 60 mm/h.

Literature Cited

1.   L. G. Merkulov, Akust. Zhurn. 5(4):432-39 (1959).

TWO TYPES OF SKELETAL CRYSTALS

S. A. Stroitelev

According to Shafranovskii [11], crystals in which, owing to a sharp difference in growth rates in different directions, a substance fills only part of the framework of a polyhedron, should be called skeletal crystals. Branched forms of skeletal crystals are commonly called dendrites. Mokievskii and Semenok [2] proposed that only forms in which the branches were curved and not parallel should be termed dendrites. The term "dendrite," however, has become firmly established in the literature, and most investigators call branching single crystals dendrites. In contrast to these, skeletal forms having polyhedral conical depressions instead of faces, in which the crystals as a whole forms a continuous skeleton or framework, may suitably be called framework, or skeletal, crystals.

Skeletal crystals are formed most frequently on crystallization from supercooled melts or heavily supersaturated solutions as well as on the electrolytic precipitation of metals.

A summary of the various views as to the reasons underlying the formation of skeletal crystals (including dendrites) is contained in the groups of replies to an international questionnaire in the journal "Metallurg" [1] and in monographs by Shafranovskii [11] and Saratovkin [3] Despite much study, the causes responsible for the formation of skeletal crystals have not been finally established.

Many workers, following Leman's example, associate the formation of skeletal forms only with the nonuniform access of material to different parts of the crystal, citing the greater accessibility of its vertices and edges as compared with the middle of the faces [3, 10, 11]. This feature of crystals, however, is constant, while skeletal forms arise only under certain conditions. Depending on the state of the crystallization medium, the form of growth of the crystals, and hence also the faces, vertices, and edges, vary in a regular way. The crystallographic direction of the branches of dendrites also does not remain constant. For example, in NH_4Cl, Saratovkin [3] established "octahedral dendrites" in which the dominant growth occurred in the [100] direction, and "cubic dendrites," the branches of which, in the presence of impurity, grew in the [111] direction. Different crystallographic directions are observed in the axes of copper, silver, and other dendrites. All this shows that the causes of dendritic crystallization lie not only in the way in which material reaches different parts of the crystals but also in its nonuniform assimilation in various crystallographic directions.

Skeletal crystals represent one form of crystal growth. Their formation must therefore be considered from the same standpoint as that of any other growth form.

Our studies on the production of crystals of assigned form showed that the relative growth rates of different faces depended on their structure and the state of the atoms in the crystallization medium [4, 5]. Important factors here include the degree of dissociation of the compound, the relative concentrations of its components, their chemical activity, and the strength of the bond with co-dissolved materials. From such standpoints we can explain the effect of impurities [4], supersaturation [7], and other factors on the form of growth [4, 7] and the form of dissolution figures [6] of the crystals.

In the crystal lattices of compounds we may distinguish those directions in which atoms of different types are joined, i.e., directions with a heteropolar bond. In addition to these, we may distinguish planes consisting of atoms of one sort, i.e., directions with a homopolar bond. It is established experimentally that, during crystal growth, the growth rate varies in different ways, depending on the crystallization conditions in various directions. This is of very great importance in the formation of the habit in crystals.

Fig. 1. Plane dendrites of leboite.

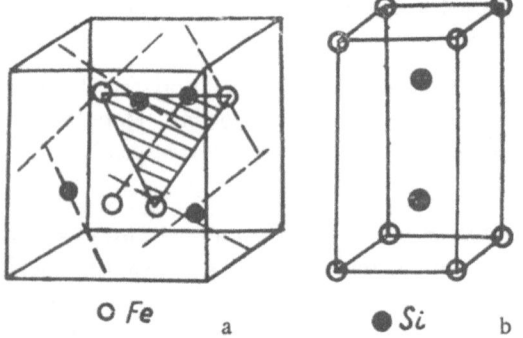

O Fe   a   ● Si   b

Fig. 2. Structures of (a) iron monosilicide and (b) leboite.

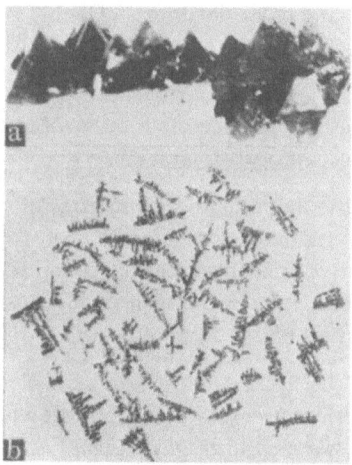

Fig. 3. a) Branched dendrites of FeSi broken up into
tetrahedra, and b) branched dendrite of leboite.

The results of our experiments illustrate this importance of the growth rate; it follows from them that, when the crystallization medium contains an excess of one of the components, the inter-metallic semiconducting compounds GaAs, InAs, GaSb, InSb, AlSb form dendrites which break up into individual segments or lamellar twins. The smallest growth rate in this case belongs to the tetrahedral faces perpendicular to the crystal direction with the heteropolar bond. Such dendrites are formed in gallium arsenide even when crystallized from solution in zinc, silver, or tin melts [5]. Such lamellar crystals and dendrites were found in leboite ($FeSi_2$) [8]. We found that one generation of leboite precipitated from solution in tin melt earlier than the crystals of free silicon, and then skeletal crystals and lamellas of this ferrosilicide with the pinacoid faces (001) most developed (Fig. 1) were formed. Another generation precipitated later than the silicon crystals. In this case three-dimensional dendrites developed, the axes of these being directed at 90°. Parallel to the (001) faces in the crystal lattice of leboite, single planes composed of iron atoms and two composed of silicon (Fig. 2) alternate.

The latter shows that the silicon, precipitating later and found in solution during the growth of the lamellar crystals, played the role of an excess component, as in the above-mentioned case. The dendrites, consisting of tetrahedra set parallel to one another, form iron monosilicide (FeSe) if the ratio of the two components in the melt is not 1:1 (Fig. 3). Here the most developed tetrahedral faces (111) are parallel to the crystal-lattice planes consisting of atoms of the same kind (Fig. 2).

Lamellar dendrites of cadmium iodide (CdI_2) are formed from pure aqueous solution even for weak supersaturations (Fig. 4). The lowest growth rate here belongs to the pinacoid faces (0001), perpendicular to the direction with the heteropolar bond, in which planes consisting of Cd ions alternate with two planes composed of I ions.

It is known that the hydration energy of iodine ions is 62 and that of cadmium ions 445 kcal/g·ion. Hence the cadmium ions bind themselves to water molecules in solution seven times more strongly than iodine ions, and have correspondingly

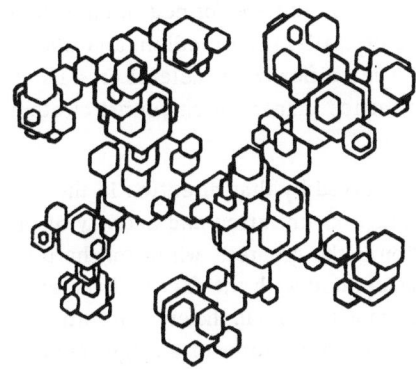

Fig. 4. Lamellar dendrites of CdI_2 (after Saratovkin).

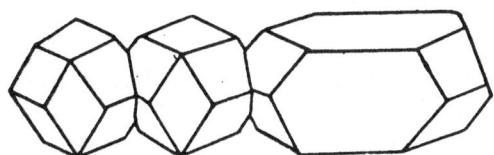

Fig. 5. Branch of a FeSi dendrite broken up into rhombic dodecahedra.

smaller activity toward the formation of lattice planes parallel to the pinacoid.

The skeletal crystals of GaAs, $FeSi_2$, FeSi, and CdI_2 mentioned are characterized by the fact that the most developed of their faces are parallel to the planes consisting of atoms of the same type. The number of examples could be greatly increased. But even from those mentioned we can clearly picture the special type of skeletal crystals formed as a result of the low growth rate of those crystal directions in which planes consisting of one kind of atom alternate. Skeletal crystals of this kind are formed because in the crystallization medium one of the components has a greater concentration or activity than the other. The crystals in question, peculiar from both the genetic and the crystallochemical points of view, may be referred to as skeletal crystals of the first kind.

The situation is different when the concentration ratio of the components in the crystallization medium corresponds to that in the actual compound, and the chemical activity of the different ions to the elements of the solution are the same. Experiment shows that, if the quantitative ratio of gallium to arsenic in the crystallization medium equals 1:1, gallium arsenide dendrites of another type result. In this case the [111] direction, in which the heteropolar bond occurs, has the greater growth rate, and the most developed faces are the (110), (211), and (hkl) parallel to the planes composed of atoms of both types. The individual branches of the dendrite have the form of needles with hexagonal or triangular cross sections [5].

To the second type of skeletal crystals also belong the three-dimensional leboite dendrites, the branches of which are directed along the [100], [010], and [001], where planes consisting of atoms of one kind alternate (Fig. 3). The precipitation of such dendrites later than the silicon crystals indicates that at this period the ratio between the concentrations of the Fe and Si atoms in solution was the same as in $FeSi_2$, i.e., 1:2. Hence the formation of lattice planes composed of silicon atoms and those composed of iron atoms was equally possible The result was that the crystal growth rate in the [001] direction, which was the lowest when there was an excess of silicon atoms in solution, now became the greatest.

If the ratio of Fe and Si in the Fe—Si melt is equal to 1:1, iron monosilicide dendrites with faces of rhombic dodecahedra (110) parallel to the lattice planes composed of different kinds of atoms* (see Fig. 5) are formed.

Ammonium chloride usually precipitates from pure aqueous solutions in the form of dendrites. The greatest growth rate here belongs to the cube faces (100) parallel to the crystal-lattice planes composed of ions of one sort. The hydration energy of the chlorine and ammonium ions are identical and equal to 79 kcal/g · ion. This indicates that the interaction of the different ions with the solvent will be the same, as will be their capacity for forming crystal lattice planes. If, however, we introduce urea or metal chlorides (both of which contain ions in common with NH_4Cl) into solution, then together with the dendrites normal crystals or dendrites with another direction of the axes [10] precipitate.

*We judged the amount of Si and Fe in the ferrosilicon from its specific gravity. For specific gravity 5.88, ferrosilicon contains 50% Fe and 50% Si.

It follows from what has been said that skeletal crystals of the second kind are formed in cases when a stoichiometric ratio of the components of the compound occurs in the crystallization medium and when their activities with respect to the co-dissolved substances are the same. They differ from skeletal crystals of the first kind in that their most highly developed faces correspond to the lattice planes which consist of atoms of one or the other component.

In the formation of skeletal crystals, therefore, a special part is played by that direction in the crystal in which planes composed of atoms of one kind alternate. For the same concentration and activity of the components in the crystallization medium, the growth rate in this direction is the greatest, while for sharply differing concentrations and activities it is the smallest. All of which is associated with the fact that the excess component, surrounding the particles of the other component and interacting with them in the crystallization medium, hinders their attachment to the faces parallel to the lattice planes composed of atoms of a single kind, and aids their attachment to planes composed of different atoms [4, 5]. For stoichiometric ratio and identical activity of the components, the formation of lattice planes by one type of atom or the other is equally possible. The chemical bond is strongest in the direction in which different types of atoms are joined, and hence under these conditions it is here that the crystal growth rate is greatest.

It should also be noted that normal crystals may also be divided into different types according to the nature of the lattice planes at the surface [9], and the general classification may be presented as follows:

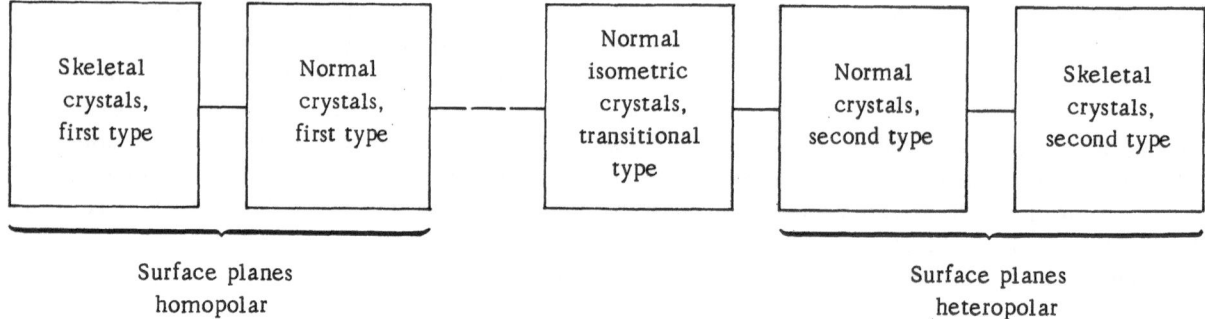

The crystals of simple substances deserve special study; we may only note here that they possess great morphological and genetic similarity to the crystals of compounds.

Literature Cited

1. Questionnaire on Dendrites, Metallurg Nos. 7 and 8 (1932).
2. V. A. Mokievskii and S. N. Semenyuk, Zap. Mineral. Obshch. 81(2):(1952).
3. D. D. Saratovkin, Dendrite Crystallization (Metallurgizdat, 1957).
4. S. A. Stroitelev, Geology and Geophysics, S. O. Akad. Nauk SSSR, No. 6 (1961).
5. S. A. Stroitelev, Collection: "Crystallization and Phase Transformations." Izd. Akad. Nauk Belorus.SSR (1962).
6. S. A. Stroitelev, Form of Etch Figures of Crystals, Nauchn.-Tekhn. Inform. TsNIIO, Sheet No. 4 (1961).
7. S. A. Stroitelev, Byul. TsNIIO, No. 3 (1961).
8. S. A. Stroitelev and O. N. Loginova, Byul. TsNIIO, No. 3 (1962).
9. V. S. Sobolev, Introduction to the Mineralogy of Silicates (Izd. L'vov. Gos. Univ., 1949).
10. Yu. Ya. Til'mans, Tr. Odessk. Industr. Inst. 2(3) (1940).
11. I. I. Shafranovskii, Crystals of Minerals (Curved-Faced, Skeletal, and Granular Forms) (Gosgeolizdat, Moscow, 1961).

STRUCTURAL FEATURES OF ZONE-MELTING IRON

F. N. Tavadze, I. A. Bairamashvili, L. G. Sakvarelidze, V. Sh. Metreveli,
N. A. Zoidze, and G. V. Tsagareishvili

A number of experimental papers have been devoted to obtaining pure iron by the zone-melting method. The possibility of producing single crystals of iron during zone refining was noted in [1], and it was shown that the orientation of the single crystal is formed in the δ-region and remains unchanged down to room temperature.

In this connection, the question as to whether there is an intermediate transformation of iron into the γ-form on cooling from the δ-region, and if so by what mechanism this takes place, remains unanswered. Furthermore, there is very little information on the perfection and dislocation structure of the iron crystals obtained by zone melting. The aim of the present paper is to study the structural characteristics of iron from zone melting.

Materials and Experimental Method

The raw material for subsequent zone melting was carbonyl iron of high purity (class A2) remelted in high vacuum and cast into samples 150 to 200 mm long and 10 to 12 mm in diameter.

The remelting was effected by one of two methods of zone melting: horizontal, with high-frequency heating; or vertical, with an electron beam.

The horizontal melting was effected in an atmosphere of purified helium in CaO boats. The zone width was 10% of the total length of the sample. The rate of displacing the inductor varied from 0.6 to 1 mm/min. The number of passes was 8 to 10.

The vertical zone melting was effected in high vacuum (10^{-4} to $5 \cdot 10^{-5}$ mm Hg). The zone width was 2 to 3 mm, the rate of displacement 1 to 2 mm/min. The number of passes was 4 to 5. In order to vary the temperature gradient and to use the phenomenon of electric transport of impurities during the vertical zone melting, a direct current of density up to 400 A/cm^2 was passed through the sample.

The samples were studied by electron and optical microscopy and X-ray analysis. The microstructure of the samples was revealed by electrolytic etching in a mixture of 78 ml H_3PO_4, 10 ml H_2SO_4, 4 ml H_2O, and 8% Cr_2O_3.

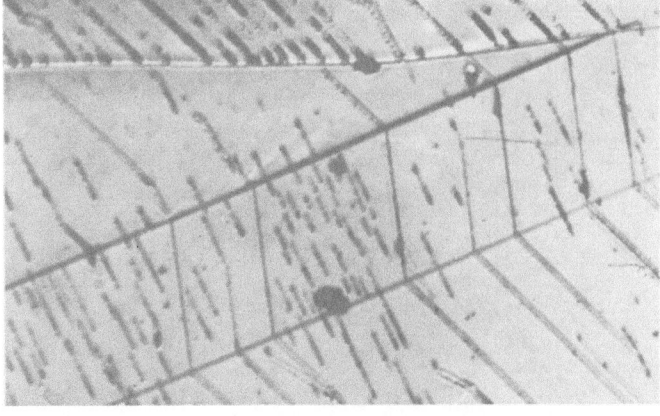

Fig. 1. Surface relief of iron. ×200.

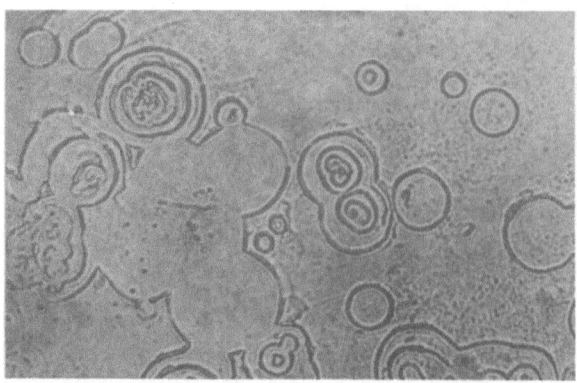

Fig. 2. Dislocation loops. × 600.

Results of the Experiments

The bars obtained by zone melting consisted of coarse (up to 30 mm) columnar crystals drawn out in the direction of motion of the zone. Microstructural study showed that the crystallites contained sub-boundaries, the so-called "α veins." This "vein structure" is described in [2, 3] and is connected with the polygonziation of pure iron.

In studying the unetched surface of samples obtained by the horizontal zone-melting method, relief was found on the surfaces of the bars (Fig. 1), the structure of which indicated the presence of a coarser network, differing from the final fine network of α-iron grains.

Each grain of the coarse network contains rectilinear bands, which sometimes break, as if at a twin boundary. The bands are most frequently set at an angle of 45° to the sample axis.

X-ray determination of the orientation of individual, randomly selected crystals of α-iron failed to show any preferred growth direction.

In samples obtained by horizontal zone melting, there is a low dislocation density. The high purity of ˙ the iron and the great distance between dislocations contribute to the extremely easy activation of the Frank—Read sources. Figures 2 and 3 show a large number of activated sources. Figure 4 shows a unique photograph of a Frank—Read source in every successive stage of its development. The point etching of dislocation loops is typical.

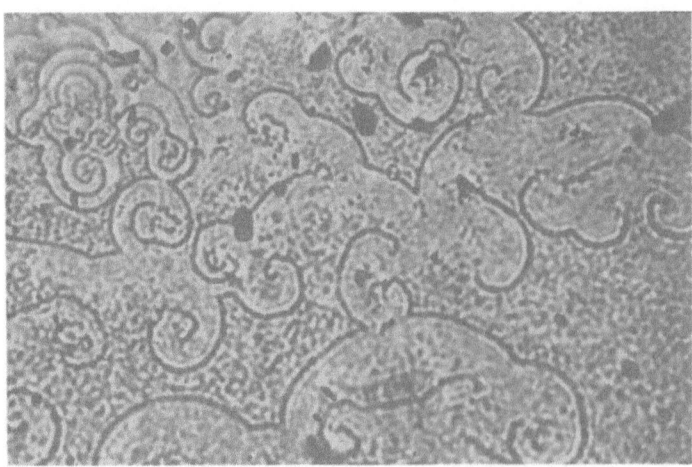

Fig. 3. Dislocation spirals. × 700.

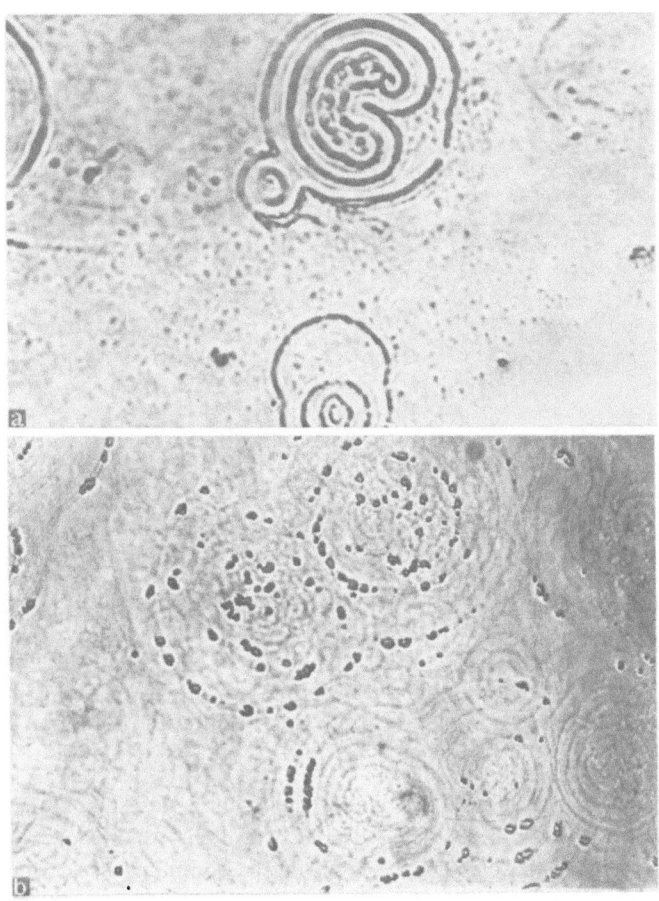

Fig. 4. Frank—Read source (a), and point etching
of loops (b). × 900.

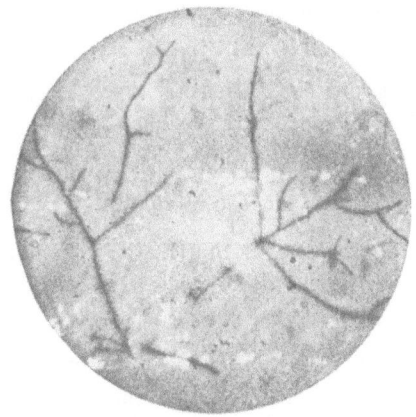

Fig. 5. Micro X-ray photograph of "vein
structure." × 250.

Figure 5 shows a micro X-ray photograph of the "vein structure" of iron obtained by reflection in iron radiation [4]. The great extinction contrast is due to the fairly high perfection of the iron crystals grown.

In samples obtained by vertical zone melting, we find bands of etch pits, the disposition of which depends on local changes in melting conditions (Fig. 6). Within each band the pits are disposed in a regular lattice, their density being $9 \cdot 10^6$ per cm^2. The density does not depend on the direction of the section. The number and disposition of the pits appear reproducible after repeated repolishings and etchings up to 30 μ. Annealing at 1200°C for 1 h and water-quenching alters neither the etchability nor the disposition of the pits.

Substantial purification from impurities, especially carbon, was not achieved in our experiments.

Discussion of Results

The development of relief on the free surface of face-centered metals in a slightly oxidizing atmosphere is described in [5, 6]. In our experiments the surface relief evidently reflected the structure of the γ-form of

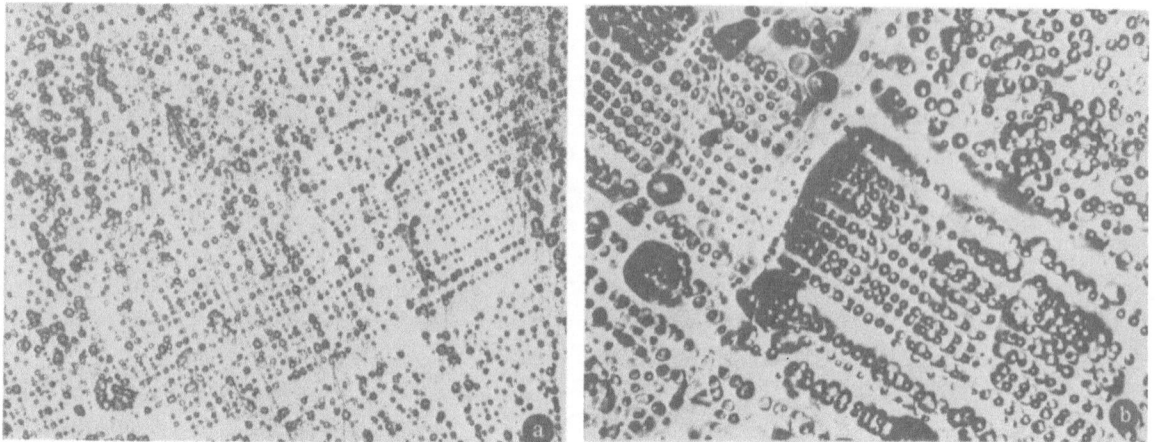

Bands of etch pits (a), and characteristic regular
disposition of etch pits (b). ×300.

iron. This is indicated by the presence of twins and the rectilinearity of the bands in the "pseudograins" of
γ-iron. The rectilinear bands are traces of the outcropping of the (111) planes [5].

Thus in the zone melting of iron there is a transformation into the α-form. By connecting the orientation
of the relief with that of the α-crystals, we may find the orientation relations of the $\gamma \rightarrow \alpha$-polymorphic
transformation.

It was noted that the traces of the (111) planes of the γ-form were most frequently set at 45° to the sample
axis. The α- crystals, however, were disposed in a disorderly fashion with respect to this axis. Hence we were not
observing a regular $\gamma \rightarrow \alpha$ transformation. But the recurring predominant disposition of the γ-crystals indicates
a regular $\delta \rightarrow \gamma$ rearrangement. Later in the $\gamma \rightarrow \alpha$ transformation this regularity may be broken owing to polygonization phenomena in the α-iron. Upon the suppression of polygonization, and the combining of our data
with the data of [1], we should expect that the $\delta \rightarrow \gamma \rightarrow \alpha$ transformation will take place in such a way that,
by the choice of certain rates, δ will be transformed into γ and along the same planes γ will transform into α,
preserving the original orientation.

If the crystallography of the $\delta \rightarrow \gamma$-polymorphic transformation was of the martensite type, i.e.,
$(110)_\delta \parallel (111)_\gamma$, in our experiments the δ grew in the [100] direction, in accordance with published data [7].

The regular disposition of etch pits and the distance between them (10^{-4} cm) obtained in the vertical-
zone-melting sample are very reminiscent of the disposition of dislocations in the lattice described by Taylor [8].

Such dislocation lattices are frequently found in very different metals and alloys: in nickel [9], titanium
alloys [10], and copper, zinc, and iron silicide [11]. In the papers cited, the development of the lattice is explained by the mechanism of "polygonization by slip."

In our case the formation of the lattice takes place in the δ-region, since it plainly depends on local
variations of the melting conditions. Despite the high temperature, the formation of the lattice must be as-
cribed to "polygonization by slip" and not to creeping, since in the latter case we should expect the dislocations to be built up into more stable dislocation walls [10]. Actually, in the unmelted parts of the sample (at
the clamps), the etch pits are disposed along the sub-boundaries of the high-temperature form (Fig. 7). Slip is
clearly caused by thermal stresses.

If the crystals grew in the [100] direction, the Burgers vector of the lattice dislocations had the same direction. The existence of a dislocation lattice with a [100] Burgers vector was shown experimentally in an
electron-microscope study of thin films of strained iron [12].

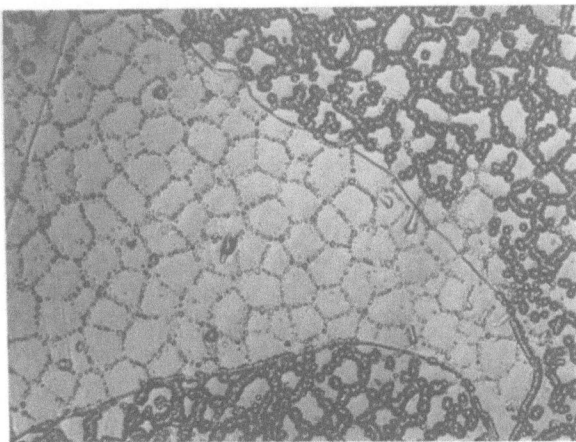

Fig. 7. Disposition of etch pits on the sub-boundaries of
the melted part of the sample. ×400.

In the samples obtained by vertical zone melting, the dislocation density was far higher than in horizontally melted samples. The vertical melting was effected at higher rates of zone displacement and lower temperature gradients. Clearly the mechanism of dislocation formation is of the vacancy type, or, since a large number of dislocations of the same sign are obtained, they originate by the mechanism described by Amelinckx[13]. Hence the parts of the vertical-melting samples free from dislocations solidified at lower crystallization-front displacement rates and lower local temperature gradients.

As mentioned earlier, heating to 1200°C and water-quenching changed neither the etchability nor the disposition of the etch pits. Evidently the etching reveals submicroscopic pores. Electron microscopic study of the pits showed that they all had flat bottoms, i.e., did not constitute dislocation-outcrop sites, at all events at room temperature. Hence we may suppose that the pores arose by way of the deposition of vacancies at the dislocation lattice formed in the high-temperature form of the iron. Subsequently, the dislocation lattice is disrupted, and we recognize it only from the regular disposition of the remaining micropores. It is difficult to determine at exactly what point the dislocation lattice breaks up (at the time of the polymorphic transformations or on polygonization of the α-iron). It is thus hard to decide whether there is a transfer of dislocation defects from one form of iron to another. Such a conclusion may be drawn after a single crystal is obtained.

As regards the possibility of fixing different positions of one and the same dislocation loop generated by a Frank—Read source (Fig. 2), this has been considered by Amelinckx [13]. The regular intervals between the decoration sites of the dislocation may be explained in the following way: Curving out from the sources, the dislocation gathers around itself a Cottrell atmosphere, together with which it proceeds to move. Hence as time passes the stress required for the displacement of the loop increases. At a critical stress the loop breaks free from the atmosphere, which it leaves behind, as the atmosphere is not capable of being drawn apart. The freed dislocation rapidly moves into new regions, where it is again retarded, and so on. The impurities meanwhile manage to redistribute themselves along the dislocation tube with the precipitation of a phase at the moment when the dislocation breaks off, as clearly seen from the looped disposition of the etch points.

The mechanism described is in good agreement with the ideas of Fisher, et al., as to the role of the migration of a dislocation and grain boundaries in precipitation processes [14].

Literature Cited

1.   H. Hillman and A. Mager, Arch. Eisenhüttenwes. 1:663 (1960).
2.   H. Hahnemann, Arch. Eisenhüttenwes. 6:507 (1932-33).

3. E. Ammermann and U. Kornfeld, Stahl und Eisen 49:1192 (1929).

4. G. S. Barrett, Trans. Amer. Inst. Min. (Metall) Engrs. 161:15 (1945).

5. I. Moreau and I. Benard, J. Inst. Met. 83:87 (1955).

6. E. Votava and A. Berghezan, Acta Met. 7:6 (1959).

7. W. M. Martius, Advances in the Physics of Metals, Vol. II [Russian translation] (Metallurgizdat, 1952).

8. G. I. Taylor, Proc. Roy. Soc. 145A:262 (1934).

9. P. Feltham, Inst. Met. 86:237 (1957-58).

10. T. H. Schofield and A. E. Bacon, Acta Metallurgica 7(6):(1956).

11. C. G. Dunn and W. R. Hibbard, Acta Metallurgica 3:409 (1955).

12. P. B. Hirsch, J. Inst. of Metals (August, 1959).

13. S. Amelinckx, Acta Metallurgica 6(1):34 (1958).

14. A. H. Cottrell, Structure and Formation of Metals and Alloys [Russian translation] (Metallurgizdat, 1961).

PHASE TRANSFORMATIONS IN THE PROCESSES OF REDUCING
URANIUM OXIDES

V. M. Zhukovskii, E. V. Tkachenko, V. G. Vlasov, and V. N. Strekalovskii

The study of phase transformations in processes of the "solid—gas" and "solid—solid" type, for example, in the reduction of metal oxides by various reducing agents, is of considerable interest for both metallurgists and crystallochemists. Considerable attention must be paid to structural features of the oxide phases which arise, the nature of their defects, and the character of the semiconducting properties. In this connection it is useful to compare experimental data on the kinetics of the reduction of metal oxides with the structural character- istics of intermediate products, devoting special consideration to the formation of nonstoichiometric compounds. Such comparisons, in our view, are a necessary step on the way to setting up special investigations into the ther- modynamics and kinetics of transformations in nonstoichiometric compounds.

In particular, it would seem extremely interesting to study such processes in the uranium—oxygen sys- tem, which contains a large number of phases of constant composition as well as a wide region of solid solu- tions and nonstoichiometric compounds [1-7]. What we have said may be illustrated by a comparison of the

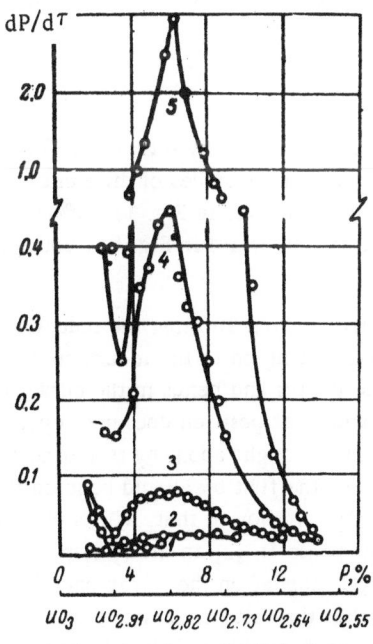

Fig. 1. Isotherms of the reduction of UO_3 by solid carbon (P = degree of reduction, %): 1) temperature 400°C; 2) 425; 3) 450; 4) 475; 5) 500.

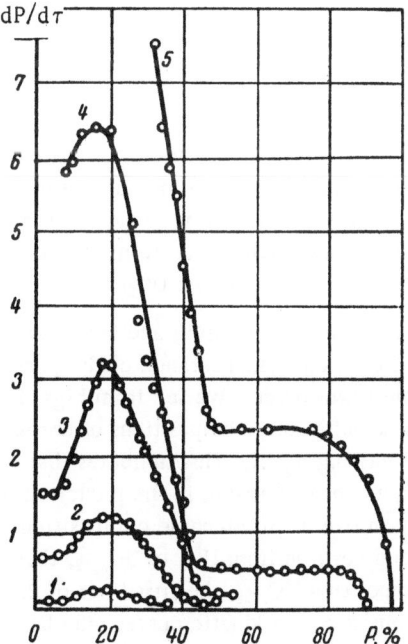

Fig. 2. Isotherms of the reduction of UO_3 by dissociated ammonia. $P_{3H_2} + N_2$ = 200 mm Hg: 1) temperature 350°C; 2) 400; 3) 425; 4) 450; 5) 500.

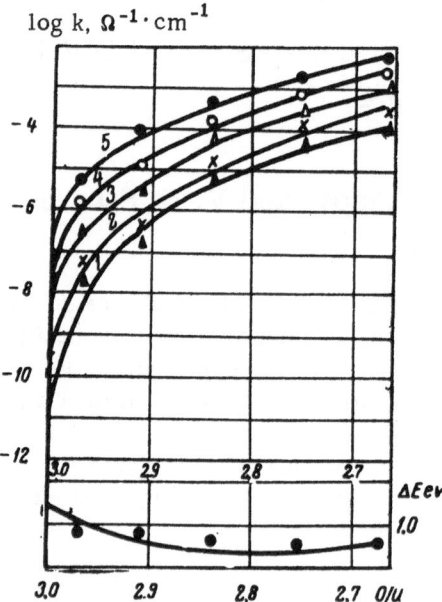

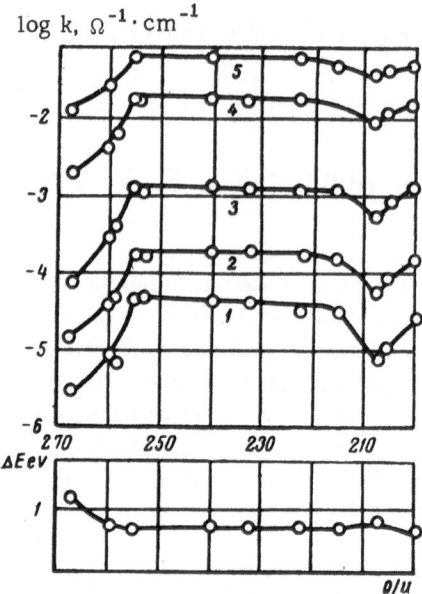

Fig. 3. Isotherms of the electrical conductivity of uranium oxides in the composition range UO_3 to U_3O_8: 1) temperature 25°C; 2) 50; 3) 100; 4) 150; 5) 200.

Fig. 4. Isotherms of the electrical conductivity of uranium oxides in the composition range U_3O_8 to UO_2: 1) temperature 25°C; 2) 100; 3) 200; 4) 400; 5) 600.

results of kinetic and X-ray studies of the phase transformations of uranium oxides by solid (Fig. 1) and gaseous
(Fig. 2) reducing agents with measurements of electrical conductivity taken for oxides of different compositions
(Figs. 3 and 4) and also during reduction processes (Fig. 5). In plotting the curves in Fig. 1, 100% reduction
corresponds to a complete transformation of uranium trioxide into the metal, while in Fig. 2 it corresponds to
the transformation of UO_3 and UO_2. *

   It is known that minima and bends on the kinetic curves obtained when examining the reduction of metal
oxides arise on the disappearance of the phase richer in oxygen and on passing on to the reduction of the corresponding lower oxide. We see from Figs. 1 and 2 that characteristic breaks and bends in the curves take place
for quite definite total compositions of the solid reduction products, and their position does not depend on the
form of reducing agent. This indicates that in the first instance they are brought about by structural features
of the crystal phases arising in the reduction process. Thus in both cases the first break and bend on the curves
occur for an overall solid-phase composition $UO_{2.92}$. X-ray phase analysis† showed that, with a change in the
overall composition from UO_3 to $UO_{2.92}$, there was a single amorphous phase UO_{3-p} in the system. Beginning
from composition $UO_{2.92}$, the first lines of the rhombic structure of U_3O_8 appear on the X-ray photographs. Hence
the first break on the kinetic curves may be associated with the fact that the last portions of oxygen pass out
of the solid solution based on uranium trioxide with some difficulty, and also with certain difficulties in the
appearance of nuclei with rhombic structure of U_3O_8. Hence it corresponds to the lower limit of the nonstoichiometric uranium trioxide UO_{3-p}.

---

*The method of the kinetic studies, the types of apparatus, and analysis of the experimental data are described
 in [8-11].
† The X-ray examination was made as in [12-15].

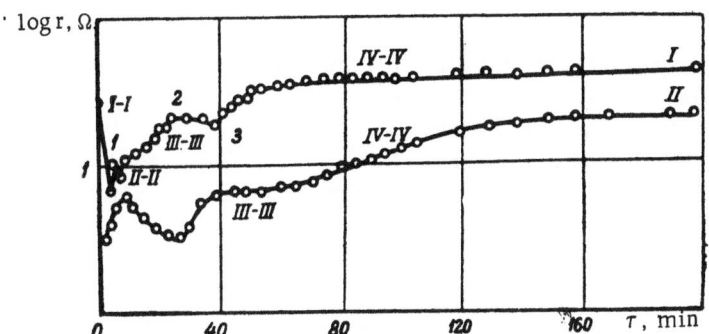

Fig. 5. Variation of resistance R with time τ in the reduction of (I) UO_3 and (II) U_3O_8 by hydrogen at 450°C [17].

Figure 3 shows isotherms of the specific electrical conductivity in the composition range UO_3 to $UO_{2.67}$.* From this figure we see that the electrical conductivities of UO_3 and U_3O_8 differ substantially (by about seven orders). Slight reduction in the oxygen content of nonstoichiometric uranium trioxide produces a sharp increase in electrical conductivity. Thus, changing the composition from UO_3 to $UO_{2.97}$ leads to an increase in conductivity of four orders. This is apparently associated with the increased defectiveness of the uranium trioxide structure as oxygen is removed from it.

The next characteristic break in the kinetic curves of Figs. 1 and 2 occurs for a total solid-phase composition which corresponds approximately to the oxide $UO_{2.56}$. X-ray phase analysis showed that, at roughly this solid-phase composition, the rhombic U_3O_8 lines on the X-ray photographs were joined by the first lines of cubic structure. Thus this bend is due to difficulties in separating the last portions of oxygen from solid solutions based on the mixed oxide and in forming nuclei of the new phase. It gives the lower limit to the non-stoichiometric mixed oxide of uranium ($U_3O_{8-z_{max}}$). It should be noted that for the same solid-phase composition there is a bend on the isotherms of electrical conductivity, as shown in Fig. 4. We also see that, with a reduction in the oxygen content in the mixed oxide of uranium from $UO_{2.67}$ to $UO_{2.56}$, the electrical conductivity increases by more than one order. In this case also, the rise in conductivity is evidently due to the increased defectiveness of the rhombic structure of uranium mixed oxide.

The third characteristic bend is found at a solid-phase composition corresponding to the oxide $UO_{2.25}$. There is also a sharp bend in the electrical-conductivity isotherms (Fig. 4) for this composition: On reducing the oxygen content (below the oxide $UO_{2.25}$) the conductivity falls, and for a composition $UO_{2.07}$ passes through a minimum, which agrees with published data on the upper limit of the single-phase region $UO_{2+x_{max}}$ in the temperature range 500 to 600°C: $0.10 > x_{max} > 0.06$ [1, 2, 6].

Thus the kinetic studies, the x-ray phase-analysis data, and the electrical-conductivity measurements show that the phase transformations occurring during the reduction of uranium oxides are as follows [11-15, 17]:

$$UO_3 \rightarrow UO_{3-p_{max}} \rightarrow U_3O_8 \rightarrow U_3O_{8-z_{max}} \rightarrow U_4O_9 \rightarrow UO_{2+x_{max}} \rightarrow UO_{2+x}.$$

We must emphasize the value of further examination of the electrical conductivities of solid oxides. In particular, this method has enabled us to reproduce the pattern of phase transformations which take place during the reduction of the higher oxides of uranium. This proved identical with those established by means of kinetic and X-ray methods. Similarly, we recognize the possibility, in principle, of using the electrical-conductivity method for studying reduction interactions of the "solid—gas" and "solid—solid" types. Attractive here are the simplicity and practicability of the method, as well as its high sensitivity to slight structural disruptions and deviations from stoichiometry.

---

* The electrical-conductivity measurements were made as described in [12, 13] and [16, 17].

Literature Cited

1. H. Hoekstra and S. Siegel, Transactions of the First International Conference on the Peaceful Uses of Atomic Energy [Russian translation] 7:482 (1958).
2. F. Gronvold, J. Inorg. Nucl. Chem. 1:357 (1955).
3. R. K. Willard, J. W. Moody, and H. L. Goering, J. Inorg. Nucl. Chem. 6(1):19 (1958).
4. G. Belle, Transactions of the First International Conference on the Peaceful Uses of Atomic Energy [Russian translation] 6:231 (1959).
5. Atomnaya énergiya 4(2):215 (1958).
6. B. E. Schaner, J. Nucl. Mater. 2(2):110 (1960).
7. H. Hoekstra, A. Santaro, and S. Siegel, J. Inorg. Nucl. Chem. 18(1):166 (1961).
8. V. G. Vlasov and V. A. Kozlov, Zh. Prikl. Khim. 33(4):760 (1960).
9. V. N. Strekalovskii and V. G. Vlasov, Zh. Prikl. Khim. 34(1):38 (1961).
10. V. G. Vlasov and V. N. Shalaginov, Zh. Prikl. Khim. 34(1):20 (1961).
11. V. G. Vlasov and V. M. Zhukovskii, Kinetika i Kataliz 3(6):882 (1962).
12. V. M. Zhukovskii, V. G. Vlasov, and A. G. Lebedev, Fiz. Metal. i Metalloved., 14(2):319 (1962).
13. V. M. Zhukovskii, V. G. Vlasov, and A. G. Lebedev, Fiz. Metal. i Metalloved. 14(3):475 (1962).
14. V. M. Zhukovskii, E. V. Tkachenko, and V. G. Vlasov, Fiz. Metal. i Metalloved. 15(2):210 (1963).
15. E. V. Tkachenko and V. G. Vlasov, Fiz. Metal. i Metalloved. 15(2):239 (1963).
16. S. F. Pal'guev and A. D. Neuimin, Collected Works of the Ural Branch of the Academy of Sciences Institute of Electrochemistry (1):111 (1960); English translation: Consultants Bureau, 1961, p. 90.
17. V. N. Strekalovskii, A. D. Neuimin, and A. F. Bessonov, Zh. Fiz. Khim. 36(6):1355 (1962).

THERMODYNAMICS OF PHASE TRANSFORMATIONS OF THE INTERSTITIAL SOLUTION IN FROZEN SOILS AND MOUNTAIN ROCKS

N. S. Ivanov

Water in soils and mountain rocks may exist in the following states as a result of the physicochemical composition of dissolved substances and the degree of orientation of the water molecules in the force fields of the particles:

1. Free water. This category is found in coarse skeletal and crumbling mountain rocks saturated with weakly mineralized water.
2. Free solution. This category is typical of coarse-grained and crumbling mountain rocks saturated with solutions having the properties of ionic and molecular solutions and colloidal suspensions.
3. Bound or oriented water. In finely dispersed media, the water finds itself in a strong particle field, and acquires new thermodynamic properties. The most strongly oriented layers of interstitial water do not contain dissolved substances.
4. Bound interstitial solution. In view of the presence of dissolved substances, the bound water constitutes a bound solution. In these solutions the water molecules lie under the influence of the force fields not only of particles and exchange cations, but also of free ions.

We shall consider the equilibrium conditions of interstitial water in frozen soils and mountain rocks in accordance with the categories thus distinguished.

Cryogenic Phase Transformations of Free Interstitial Water

The main thermodynamic conditions of the phase equilibrium of free interstitial water in the two-phase ice—water system are the equations of pressures, temperatures, and chemical potentials:

$$\left.\begin{array}{l} T_s = T_l = T \\ p_s = p_l = p \\ \mu_s\,(p, T) = \mu_l\,(p, T) \end{array}\right\}. \tag{1}$$

The indices "s" and "l" refer to the solid and the liquid phases of water. Differentiating the last equation, we obtain

$$\frac{\partial\,(\mu_l - \mu_s)_p}{\partial T}\,dT + \frac{\partial\,(\mu_l - \mu_s)_T}{\partial p}\,dp = 0. \tag{2}$$

After substituting the values of the partial derivatives,

$$\left(\frac{\partial \mu_l}{\partial T}\right)_p = -s_l\,; \quad \left(\frac{\partial \mu_l}{\partial p}\right)_s = v_l\,,$$

$$\left(\frac{\partial \mu_s}{\partial T}\right)_p = -s_s\,; \quad \left(\frac{\partial \mu_s}{\partial p}\right)_s = v_s\,,$$

where s and v are referred to one molecule, we arrive at the fundamental phase-equilibrium equation for free interstitial water:

$$-\frac{q_{\mathrm{m}}}{T}\,dT + (v_l - v_{\mathrm{s}})\,dp = 0. \tag{3}$$

This equation may be written in the form

$$\frac{dp}{dT} = \frac{q_1}{T\,(v_l - v_{\mathrm{s}})} = \frac{q_0}{T\,(V_l - V_{\mathrm{s}})}\,, \tag{4}$$

which is known as the Clapeyron–Clausius equation.

The Clapeyron–Clausius equation enables us to establish a connection between the temperature and the pressure of phase transformations.

From the data of Planck [11], at $T = 273°K$, $V_l = 1000\ cm^3$, $V_{\mathrm{s}} = 1091\ cm^3$, $q_0 = 80$ cal/g, though from more precise data $q_0 = 79.69$ cal/g.

On the basis of these data we obtain

$$dT/dp = -0.0075 \quad \text{deg/atm},$$
$$dp/dT = -133 \quad \text{atm/deg}. \tag{5}$$

The Clapeyron–Clausius equation enables us to calculate the decrease in the freezing point of interstitial water in deep layers of the Earth's crust. Thus for a mean volumetric weight of mountain rocks $\gamma = 2 \cdot 10^3$ kg/m^3 we have the following relation between the freezing point and the depth of the layer:

$$\theta_{\mathrm{bf}} = -1.5 \cdot 10^{-3} \quad h, \text{ deg}, \tag{6}$$

where h is the depth of the layer in meters.

It follows from this relation that for depths of 400 to 500 m the freezing point of free interstitial water is 0.6 to 0.7°C below zero.

For sublimation–crystallization processes, Eq. (4) has the form

$$\frac{dp}{dT} = \frac{l_0}{T\,(V_{\mathrm{v}} - V_{\mathrm{s}})}\,, \tag{7}$$

where l_0 is the heat of crystallization–sublimation, and V_{v} and V_{s} are the specific volumes of vapor and ice.

TABLE 1. Temperature and Pressure at the Critical Points of H_2O

Phases	Temp., °C	Pressure
Ice I, liquid, vapor	0.0075	4.579 mm Hg
Ice I, liquid, vapor III	−22	2115 atm
Ice III, liquid, vapor V	−17	3530 atm
Ice V, liquid, vapor VI	+ 0.16	6380 atm
Ice VI, liquid, vapor VII	+81.6	21700 atm
Ice I, III, V	−24.3	3510 atm

From the data of Barnes and Weypond [2], on slow sublimation $l_0 = 689$ to 700 kcal/kg, which corresponds to the sum of the heats of fusion and vaporization, and for rapid sublimation $l_0 = 597$ to 615 kcal/kg, which is only 1.6% above the heat of vaporization. This may be associated with the vaporization of molecular complexes, the subsequent decomposition of which takes place with additional heat absorption.

The triple point of the phase equilibrium of the vapor, liquid, and solid phases of ice I are characterized by the parameters $p = 4.579$ mm Hg; $\theta = 0.0075°C$.

Apart from the ice-I modification, Tamman and Bridgeman established the existence of other forms of ice: II, III, IV, V, VI, and VII. Allowing for these forms, the phase diagram of water is as in Fig. 1 [15, 16].

This diagram shows the triple points of water for all the forms of ice except ice IV, the phase-equilibrium conditions of which have still not been established. Table 1 gives the parameters of the triple points.

Ice VII can exist at pressures of 20 to 50 katm up to 200°C. The densities of the above forms of ice are: ice I — 0.9168 to 0.9483 g/cm³; ice III — 1.15 g/cm³; ice V — 1.25 g/cm³; ice VI— 1.33 to 1.36 g/cm³; ice VII — 1.67 g/cm³.

The water which fills all natural reservoirs, as well as free interstitial water, exists at temperatures and pressures which exlude all forms of ice except ice I. A different appraisal may be made regarding oriented layers of water and water subjected to the action of artificial high-pressure fields. Such fields may, for example, be created by high-voltage discharges in liquids, i.e., the electrohydraulic effect. Investigations into the possibility of creating stable forms of ice at normal temperatures and pressures are of great theoretical and practical interest.

One of the parameters of the phase equilibrium of free interstitial water is its supercooling temperature. From the data of Bozhenova [1], the minimum supercooling in smallish samples of disrupted structure reaches −5°. By the photothermograph method we established that supercooling also occurs in mountain rocks of natural structure, reaching several degrees.

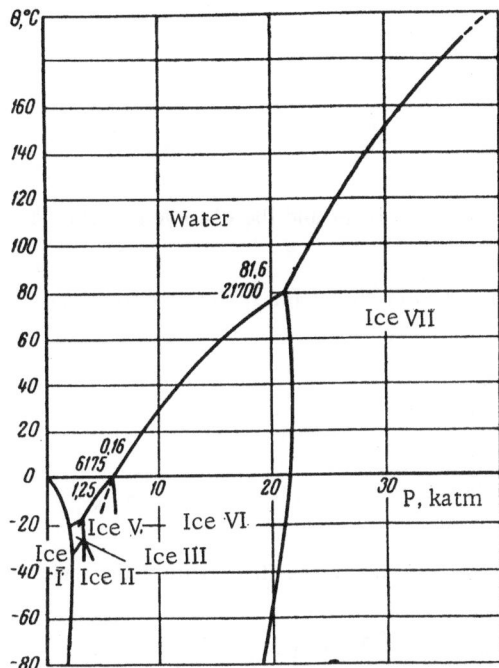

Fig. 1. Phase diagram of H_2O.

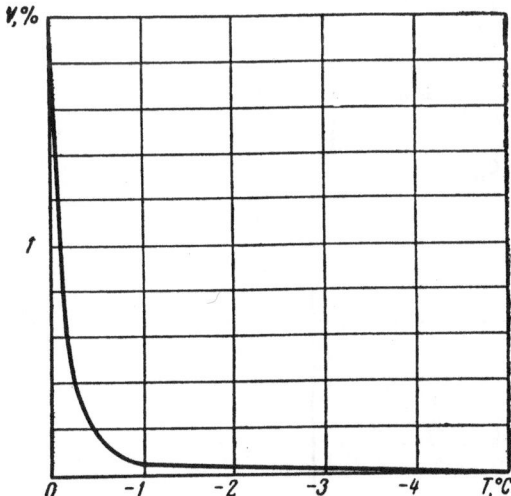

Fig. 2. Temperature-dependence of the amount of liquid phase in ice.

Cryogenic Phase Transformations of a Free Interstitial Solution

The conditions of thermodynamic equilibrium of a free interstitial solution are described by the system of equations

$$\left.\begin{aligned}
T_l &= T_s \\
p_l &= p_s \\
\mu_{wl} &= \mu_{ws} \\
\mu_{xl} &= \mu_{xs}
\end{aligned}\right\}, \tag{8}$$

where μ_{wl}, μ_{ws}, μ_{xl}, and μ_{xs} are the chemical potentials of the aqueous solvent and the dissolved substance in the liquid and solid phases.

The chemical potentials of the aqueous solvent in the liquid and solid phases for molecular solutions equal [5]

$$\begin{aligned}
\mu_{wl} &= \mu_{owl} - kTx_l , \\
\mu_{ws} &= \mu_{ows} - kTx_s ,
\end{aligned} \tag{9}$$

where μ_{owl} and μ_{ows} are the chemical potentials of pure water in the liquid and solid phases, x_l and x_s are the concentrations of dissolved substance, and k is Boltzmann's constant.

Let us expand μ_{owl} and μ_{ows} in powers of Δp and ΔT and limit ourselves to the first terms of the series:

$$\left(\frac{\partial \mu_{owl}}{\partial T}\right)_p \Delta T + \left(\frac{\partial \mu_{owl}}{\partial T}\right)_s \Delta p - x_l kT = \left(\frac{\partial \mu_{ows}}{\partial T}\right)_p \Delta T + \left(\frac{\partial \mu_{ows}}{\partial p}\right)_s \Delta p - x_s kT. \tag{10}$$

Using the thermodynamically well-known relations

$$\left.\begin{aligned}
\left(\frac{\partial \mu_{owl}}{\partial T}\right)_p = -s_l \qquad & \left(\frac{\partial \mu_{ows}}{\partial T}\right)_p = -s_s \\
\left(\frac{\partial \mu_{owl}}{\partial p}\right)_s = v_l \qquad & \left(\frac{\partial \mu_{ows}}{\partial p}\right)_s = v_s \\
(s_l - s_s)\, T = q_m &
\end{aligned}\right\}, \tag{11}$$

We obtain the fundamental equation of phase equilibrium:

$$\left.\begin{aligned}
-\frac{q}{T}\,\Delta T + (v_l - v_s)\,\Delta p = (x_l - x_s)\,kT \\[2ex]
\frac{\Delta p}{\Delta T} = \frac{q}{(v_l - v_s)\,T} + \frac{(x_l - x_s)}{(v_l - v_s)}\frac{kT}{\Delta T}
\end{aligned}\right\}. \tag{12}$$

or

For $x_l = x_s$, and in particular for $x_l = x_s = 0$, Eq. (12) becomes identical with the Clapeyron–Clausius equation.

For ionic solutions, the chemical potential of the aqueous solvent is characterized by the additional term [4]

$$\Delta\mu = \frac{1}{3}\sqrt{\frac{8\pi}{\varepsilon^3}\,\frac{\rho_l\, q^6}{T}\, x^3} , \tag{13}$$

where ε is the dielectric constant of water, q is the charge on the ion, and ρ_l is the density of the solvent (number of molecules/cm^3).

Allowing for this extra term, the phase-equilibrium equation takes the following form:

$$-\frac{q_m}{T}\Delta T + (v_l - v_s)\Delta p = (x_l - x_s)kT - \frac{1}{3}\sqrt{\frac{8\pi q^6}{\varepsilon^3 T}}(\sqrt{\rho_l\, x_l^3} - \sqrt{\rho_s\, x_s^3}).$$

(14)

The presence of dissolved substances is accompanied by a decline in the freezing point of the free interstitial solution, which may be determined from Raoult's law:

$$\delta T_x = \frac{kT^2}{q_m}\sum_{i=1}^{m}(x_l - x_s)_i,$$

(15)

where m is the number of components of the dissolved substance.

The decrease in the temperature at the onset of freezing in soils and mountain rocks relative to the freezing point of pure water may vary from a fraction of a degree to as much as several degrees. The most significant fall in the temperature at the onset of freezing may arise in the presence of artificial salinization sources and on the freezing of mineralized groundwaters. Moreover, in the freezing of water films the concentration of these rises, and this is also accompanied by a decline in the freezing point.

A considerable amount of mineralized water may be contained in ice and snow cover. Figure 2 shows the variation of the amount of liquid phase in ice with temperature [15]. The solution fills the intercrystallite channels and cavities. In sea ice the liquid phase may reach 40% or more of the total weight of the ice. Due to the considerable concentration of sea water and the fall in the freezing point, freezing occurs in the ground deposits of the arctic seas.

Cryogenic Phase Transformations of Bound Water

In finely dispersed soils and mountain rocks, a considerable portion of the interstitial water is in the oriented state and has additional potential energy of interaction in the electric fields of particles and exchange cations. Hence phase transformations of bound water are associated with additional energy losses in overcoming this interaction.

The bound water—ice system, thermodynamically speaking, is a system consisting of phases with a varying number of particles and situated in an external field of force. Another characteristic of this system is a developed phase-separation surface, which makes it necessary to consider surface phenomena.

The general equilibrium conditions of such systems for a plane phase-separation boundary may be established as follows [5]. For equilibrium of two adjacent volumes with entropies S_1 and S_2, the total entropy of the system $S = S_1 + S_2$ has a maximum value. One of the conditions of an entropy maximum for constant total number of particles $N = N_1 + N_2$ is the relation

$$\frac{\partial S}{\partial N_1} = \frac{\partial S}{\partial N_1} + \frac{\partial S}{\partial N_2}\frac{\partial N_2}{\partial N_1} = \frac{\partial S}{\partial N_1} - \frac{\partial S}{\partial N_2} = 0.$$

(16)

Remembering the thermodynamic equations

$$\left(\frac{\partial S}{\partial N_1}\right)_{U,t} = \frac{\mu_1}{T},$$

$$\left(\frac{\partial S}{\partial N_2}\right)_{U,t} = \frac{\mu_2}{T},$$

we obtain for the water−ice system

$$\left.\begin{aligned}\mu_1 &= \mu_l = \mu_2 = \mu_s \\ T_1 &= T_l = T_2 = T_T\end{aligned}\right\}. \tag{17}$$

In the absence of surface effects, system (17) is supplemented by the equation of the equality of pressures.

The chemical potentials of bound water and ice equal

$$\left.\begin{aligned}\mu_{wl}^* &= \mu_{owl} + u_l\,(x,\,y,\,z) \\ \mu_{ws}^* &= \mu_{ows} + u_s\,(x,\,y,\,z)\end{aligned}\right\}. \tag{18}$$

where μ_{owl} and μ_{ows} are the chemical potentials of free water, $u_l(x, y, z)$ and $u_s(x, y, z)$ are the potential energies of the liquid and solid phases of water in the force field, referred to one molecule.

Expanding μ_{owl} and μ_{ows}, as before, in series and keeping only the first terms of the expansions, we arrive at the phase-transformation equation of bound water:

$$-\frac{q_m}{T}\,\Delta T + (v_l - v_s)\,\Delta p + [u_l\,(x,\,y,\,z) - u_s\,(x,\,y,\,z)] = 0. \tag{19}$$

If the phase-separation boundary is not plane, there is an additional surface pressure. The condition for thermodynamic equilibrium of the phases in this case is not the equality of the volume chemical potentials of the phases, but rather the condition [4]

$$\frac{\partial \Phi_l}{\partial N} = \frac{\partial \Phi_s}{\partial N}, \tag{20}$$

where Φ_l and Φ_s are the total thermodynamic potentials of the liquid and solid phases of the water.

The total thermodynamic potentials of the phases are composed of the volume and surface potentials:

$$\left.\begin{aligned}\Phi_l &= N\mu_{wl}^* + \sigma_{ls}^*\,s \\ \Phi_s &= N\mu_{ws}^* + \sigma_{sl}^*\,s\end{aligned}\right\}. \tag{21}$$

In Eqs. (21) σ_{ls}^* and σ_{sl}^* are the surface energies of the liquid and solid phases existing in contiguity. Here we consider two surface layers at the phase-separation boundary, each of which is characterized by the interaction energy between the molecules existing at the surface of the medium and the molecules of this medium, as well as molecules of the boundary phase.

Instead of two surface layers, we may consider one layer in energy equilibrium with adjacent molecular layers of the phases. Let us assign this layer to the liquid phase. Then system (21) transforms into

$$\left.\begin{aligned}\Phi_l &= N\mu_{wl}^* + \sigma_{ls}\,s \\ \Phi_s &= N\mu_s^*\end{aligned}\right\}. \tag{22}$$

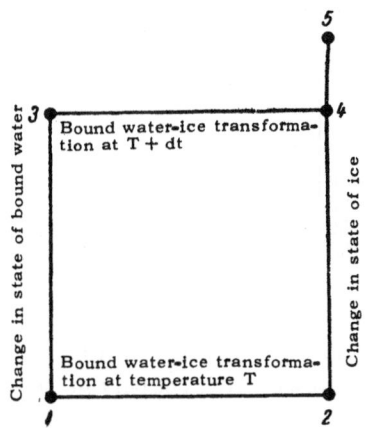

Fig. 3. Schematic representation of phase transformations of bound water in frozen soils and mountain rocks.

Substituting the values of the thermodynamic potential into Eq. (20), we find the chemical potential of the liquid phase, allowing for the Laplace pressure

$$\mu^*_{wl}(p,\ T) = \mu^*_{ws}(p,\ T) - \sigma_{ls}\frac{\partial s}{\partial N}\ . \tag{23}$$

Hence phase equilibrium for bound water with an arbitrary phase-separation surface ensues for different chemical potentials and pressures of the liquid and solid phases.

Referring to the phase equilibrium of a spherical layer of bound water covering a particle, we obtain

$$\frac{\partial s}{\partial N} = \frac{2m_m}{\rho_w r}, \tag{24}$$

where m_m is the mass of a molecule, ρ_w is the density of the water in g/cm^3, and r is the radius of the outside surface of the spherical fil

The developed expression of the phase-equilibrium conditions is then written as

$$\mu_{owl}(p,\ T) + u_l\ (x,\ y,\ z) = \mu_{ows}(p,\ T) + u_s\ (x,\ y,\ z) - \frac{2m_m\sigma_{ls}}{\rho_w r}\ . \tag{25}$$

Expanding μ_{owl} and μ_{ows} in series, we transform (25):

$$\frac{q_m - q_c}{T}\Delta T + (v_l - v_s)\Delta p = -[u_l\ (x,\ y,\ z) - u_s\ (x,\ y,\ z)] - \frac{2m_M\ \sigma_{ls}}{\rho_w r}\ , \tag{26}$$

where q_w is the heat of wetting referred to one molecule.

For free interstitial water contained in capillaries by forces of surface tension, the phase-equilibrium equation is a particular case of Eq. (26):

$$-\frac{q_m}{T}\Delta T + (v_1 - v_s)\Delta p = -\frac{2m_m\sigma_{ls}}{\rho_w r}\ . \tag{27}$$

The presence of surface tension leads to a fall in the freezing point determined by formula [16]:

$$\delta T_\sigma = \frac{\sigma_{ls}\ CT}{\rho_w q_0}\ , \tag{28}$$

where $C = 1/R_1 + 1/R_2$; R_1 and R_2 are the principal radii of curvature of the phase-separation surface, q_0 is the specific heat of fusion/solidification. The quantity δT_σ is very insignificant and in the majority of calculations need not be considered.

The fundamental quation in geocryology is the equation of the phase state of bound interstitial water. This equation is presented in the form of a relation between the unfrozen-water content W_{uw} and temperature. The principle of the equilibrium state of unfrozen-water in frozen soils and mountain rocks was first formulated in 1940 by Tsytovich [14]. The experimental basis of this principle and proof of the single-valued nature of

TABLE 2. Values of Paramters Characterizing the Freezing Resistance
of Certain Mountain Rocks

Mountain rocks	$\Delta\theta = \theta_{bf} - \theta$, °C	W_{tw} %	A	a	b
Sand (0.1 <d < 0.25)	—0.4 — — 2.4	0.100	0.106	1.406	—0.220
Sandy loam	—0.3 — —10.0	0.136	0.062	1.190	—0.056
Loam	—0.6 — — 3.2	0.218	0.182	0.656	—0.096
Jurassic clay	—0.9 — —20.0	0.330	0.235	0.233	—0.001

the relation between the bound-water content and temperature is given by Nersesova [7-9], S. Lovell [6], and others. The $W_{uw} = W_{uw}(\theta)$ relationships established by these authors are empirical and differ not only in the magnitudes of the parameters but also the nature of the functions.

We made the first attempt [3] to discover the form of this relationship for a model medium consisting of spherical particles with a cubic packing system. These constructions were based on the following scheme.

The transformation of the bound water in unit volume of a dispersed substance from the original state (1) with parameters p, T, V, into state (5) with parameters p+ dp, T + dT, V+dV may be imagined to be effected in two ways, as represented in Fig. 3.

On the path 1-3-4-5 there is a change in the temperature of the bound water to T+dT, at which a phase transformation takes place. On the path 1-2-4-5 the phase transformation of the water occurs at temperature T, after which the thermal state of the ice changes.

The increments of internal energy in the individual sections of these two transformation paths equal

$$dU_{12} = \frac{\rho_{bw} s dx}{m_{H_2O}} T \Delta S(T) + \sigma_{l_s}(T) ds - \frac{\rho_{bw} s dx}{m_{H_2O}} p_0 \Delta V_{l_s}(T) + \Delta\mu_{bw}^*(T) dN$$

$$dU_{24} = \frac{\rho_{bw} s dx}{m_{H_2O}} [c_s(T) dT - p(T) dV_s]$$

$$dU_{13} = \frac{\rho_{bw} s dx}{m_{H_2O}} [c_{bw}(T) dT - p(T) dV_{bw}] \qquad (29)$$

$$dU_{34} = \frac{\rho_{bw} s dx}{m_{H_2O}} (T + dT) \Delta S(T + dT) + \sigma_{l_s}(T + dT) ds -$$

$$\frac{\rho_{bw} s dx}{m_{H_2O}} p_0 \Delta V_{l_s}(T + dT) + \Delta\mu_w^*(T + dT) dN$$

where ρ_{bw} is the density of the bound water; s the phase-separation surface in unit volume of the dispersed substance; x the thickness of the film of bound water; S the entropy; $\Delta V_{l_s}(T)$ and $\Delta V_{l_s}(T + dT)$ the volume increments of a mole of water in the phase processes at temperatures T and T + dT; $\Delta\mu_{bw} = \mu_{wl} - \mu_{owl}$; $\mu_{bw.l}$ is the chemical potential of the bound water; c_{bw} and c_s the molar specific heats of the liquid and solid phase of the water. Making further use of the relation

$$dU_{12} + dU_{24} = dU_{13} + dU_{34}, \qquad (30)$$

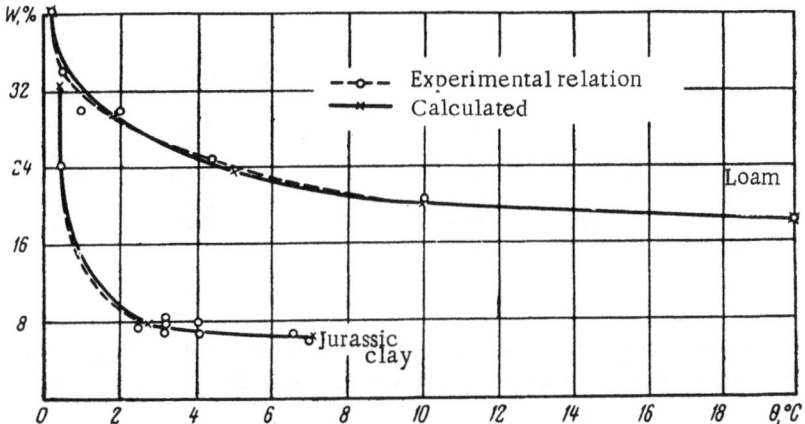

Fig. 4. Dependence of the unfrozen-water content on temperature for
certain kinds of mountain rocks.

we arrive at the equation

$$\frac{d \ln s}{dx} = \frac{p_{\text{bw}}}{\Delta \alpha \, m_{\text{H}_2\text{O}}} \left[\Delta q'_{\text{OM}}(T) - \Delta c_{ls}(T) - 2p(T) \Delta V'_{ls}(T) - p'(T) \Delta V_{ls}(T) \right].$$

(31)

Here $\Delta \alpha = \sigma_{0l}\alpha_l - \sigma_{0s}\alpha_s$, σ_{0l} and σ_{0s} are the surface tensions of the liquid and solid phases at $\theta = 0°C$, α_{il} and α_s are the temperature coefficients of the surface tensions of the liquid and solid phases, and $\Delta c_{ls} = c_l(T) - c_s(T)$.

The quantity $(d \ln s)/dx$ is determined by the structural properties of the medium, i.e., by the spectrum of particle sizes and the system of particle packing. For our model of dispersed substance we have the relation

$$s = \frac{1}{2\overline{r_0^2}} [4\overline{r_0} - \pi(\overline{r_0} + x)],$$

(32)

where $\overline{r}_0 = 2r_0/\pi$, and r_0 is the radius of a particle.

Equations (31) and (32) are basic for the derivation of the equation of the phase state of bound water in frozen soils and mountain rocks.

Having discovered the dependence of the pressure, phase-transformation heats, volume, and specific heats of the phases on temperature, from Eqs. (31) and (32) we obtain the equation of the phase state of bound waters:

$$W_{\text{uw}} = W_{\text{tw}} + A \left[\frac{1}{(1 + a \Delta \Theta + b \Delta \Theta^2)^2} - 1 \right],$$

(33)

where W_{tw} is the total water (moisture) content, and A, a, and b are parameters depending on the internal properties of the medium.

$$\Theta = \Theta - \Theta_{\text{bf}},$$

where Θ_{bf} is the temperature at the beginning of freezing.

TABLE 3. Melting Points of Ice at Various Pressures

atm	θ, °C	atm	θ, °C	atm	θ, °C
1	0.0	100	—0.75	1000	— 7.5
10	—0.075	300	—2.25	3000	—22.5
30	—0.225	500	—3.75	5000	—37.5
50	—0.375	700	—5.25	7000	—52.5
70	—0.525	900	—6.75	9000	—67.5
90	—0.675	—	—	10000	—75.0

If W_{uw} (unfrozen water) at a given temperature is potentially greater than W_{tw}, then the temperature at the beginning of freezing becomes a function of moisture content. The magnitude of the correction $\delta\Theta_x$ is found from the equation

$$\delta\Theta_x = -\frac{a}{2b} \pm \sqrt{\frac{a^2}{4b^2} - \left\{\frac{1}{b} - \left[\frac{A/b^2}{A + (W_{tw} - W_{tw}^*)}\right]^{\frac{1}{2}}\right\}}\,,$$

(34)

where W_{tw}^* is the thermodynamic equilibrium value of moisture content at the temperature of the beginning of freezing.

The values of parameters A, a, and b for various types of mountain rocks are given in Table 2.

The graphical form of the relationship $W_{uw} = W_{uw}(\Theta)$ for certain kinds of mountain rocks appears in Fig. 4. In calculating the paramters A, a, and b, we used the experimental data of Nersesova[9] et al.

Cryogenic Phase Transformations of a Bound Interstitial Solution

Interstitial water contains dissolved matter to varied degrees. Hence interstitial water, including the bound variety, has the properties of solutions.

The equation of the phase equilibrium of a bound interstitial solution, allowing for surface pressure, may be obtained by synthesizing the phase-equilibrium equations of free solutions and bound water.

Thus, for molecular bound solutions, this equation is a synthesis of Eqs. (12) and (26):

$$-\frac{q_m - q_w}{T}\Delta T + (v_l - v_s)\Delta p = (x_l - x_s)kT - [u_l(x, y, z) - u_s(x, y, z)] - \frac{2m_m\sigma_{ls}}{\rho_w r}\,.$$

(35)

For ionic solutions we correspondingly synthesize Eqs. (14) and (26):

$$-\frac{q_m - q_w}{T}\Delta T + (v_l - v_s)\Delta p = (x_l - x_s)kT - [u_l(x, y, z) - u_s(x, y, z)] -$$

$$\frac{2m_m\sigma_{ls}}{\rho r} - \frac{1}{3}\sqrt{\frac{8\pi q^6}{\varepsilon^3 T}}\left(\sqrt{\rho_l x_l^3} - \sqrt{\rho_s x_s^3}\right).$$

(36)

The phase-state equation of a bound solution in cryogenic dispersed systems, in contrast to that of bound water, allows for the fall in the freezing point in accordance with Raoult's law:

$$W_{uw} = W_{tw} + A\left[\frac{1}{[1 + a(\Delta\Theta + \delta\Theta_x + \delta\Theta_\sigma) + b(\Delta\Theta + \delta\Theta_x + \delta\Theta_\sigma)^2]^2} - 1\right],$$

(37)

where $\delta\theta_x$ and $\delta\theta_\sigma$ are the temperature drops due to the influence of the dissolved substances and the surface pressure. The analytical expression for these quantities is given by formulas (15) and (28).

Possible Forms of Ice in Frozen Soils and Mountain Rocks

The phase transformations of bound water in soils and mountain rocks take place over a wide temperature range. It is known [12] that strongly oriented layers of interstitial moisture do not freeze down to temperatures of $-78°C$. Assuming the applicability of the Clapeyron–Clausius equation to phase transformations of bound water, we may set up a table (see Table 3) characterizing the possible pressures at various phase transformation temperatures.

It follows from Table 3 that the pressure range of the phase transformations of bound water is extremely great, reaching tens of kiloatmospheres for strongly oriented layers. Comparing Table 3 with Table 1 and the phase diagram of H_2O, we envisage the possibility of different forms of ice existing in soils and mountain rocks. Such views were held, for example, by Shumskii (1955) and Parkhomenko (1956). Thus it is admissible to assume the existence of ice I, ice II, ice III, ice V, and possibly ice VI and ice VII in frozen soils and mountain rocks.

Tyutyunov [13] denied the possibility of any forms of ice save ice I existing in soils and mountain rocks. The main argument against the existence of these forms, in his view, was the absence of agreement between the energies of crystallization and the energy of forming strongly oriented layers of water. On this, Tyutyunov cited the data of Harkins on the energy of layer dehydration of solid TiO_2 particles:

Water layer	cal/mole	Water layer	cal/mole
First	16,450	Fourth	9,980
Second	11,280	Fifth	9,940
Third	10,350		

At the same time, the heat of crystallization of water at $\Theta = 0°C$ and normal pressure equals 1440 cal/mole, and the heat of vaporization at $\Theta = 100°C$ is 9,700 cal/mole.

By simple comparison of the energy of dehydration of the particles and the heats of fusion of ice at $0°C$, the existence of different forms of ice in soils and mountain rocks would appear impossible. But this comparison does not take into account the change in the structure of ice in the high-pressure region.

The heat of cryogenic phase transformations is made up of two components: the internal-energy increment, and the work associated with the change in the specific volume of water:

$$T\Delta S = \Delta U + p\,\Delta V, \tag{38}$$

where $\Delta V = V_s(p, T) - V_1(p, T)$.

For one mole of water,

$$\Delta V_m = m_{H_2O}\left(\frac{1}{\rho_s(p, T)} - \frac{1}{\rho_l(p, T)}\right)$$
$$= m_{H_2O}\frac{\rho_l(p, T) - \rho_s(p, T)}{\rho_s(p, T)\rho_l(p, T)}, \tag{39}$$

where m_{H_2O} is the mass of the gram molecules, and $\rho_l(p, T)$ and $\rho_s(p, T)$ are the densities of the phases.

TABLE 4. Energy Expended on Increasing the Volume of Water on Transforming into Various Forms of Ice

Phase transformations	Temperature of phase transformations, °C	Pressure of phase transformations, atm	Density of ice modification, g/cm³	Value of pΔV, cal/mole
Water—Ice I	+ 0	1	0.9168-0.9483	39.5-23.7
Water—Ice III	−22	2,115	1.15	$-1.22 \cdot 10^5$
Water—Ice V	−17	3,530	1.25	$-3.08 \cdot 10^5$
Water—Ice VI	+ 0.16	6,380	1.33-1.36	$-6.9-7.35 \cdot 10^5$
Water—Ice VII	+ 81.6	21,700	1.67	$-3.79 \cdot 10^6$

The densities of the liquid and solid phases are functions of temperature and pressure. The nature of these functions is still not sufficiently known. It would seem possible only to make a preliminary estimate of the molar-volume increment of bound water in phase transformations. Let us estimate, in particular, the work corresponding to the volume increment of water on transformation into various forms of ice (Table 4).

If by energy of dehydration we understand the total energy increment of water on passing from the surface of a particle into the free state, as follows from the data of Harkins quoted by Tyutyunov, the value of pΔV is quite fit to be compared with this quantity. It follows from this comparison that, even for the first layer, the energy of dehydration is an order smaller than the value of pΔV for ice III. As for the other forms of ice, for these the value of pΔV may exceed the energy of dehydration of the first layer of water molecules by two orders.

Hence, even in the energy respect, the existence of various forms of ice in soils and mountain rocks is entirely admissible. The final answer to this question, however, remains open until direct experimental proof has been obtained. It is quite possible that the presence of ions of the diffusion layer leads to a distortion in the structure of the bound water such as will prevent the formation of the crystal lattice of ice.

Literature Cited

1. A. P. Bozhenova, "Supercooling of Water on Freezing in Ground and Soils," Reports of the Frozen-Soil Research Laboratory, Collection I (Izd. Akad. Nauk SSSR, 1953).
2. B. P. Veinberg, Ice: The Properties, Development, and Vanishing of Ice. Gostekhizdat, Moscow—Leningrad, 1940).
3. N. S. Ivanov, Heat Exchange in the Cryolite Zone (Izd. Akad. Nauk SSSR, Moscow, 1962).
4. A. S. Kompaneets, Theoretical Physics (Gostekhizdat, Moscow, 1957).
5. L. Landau and E. Lifshits, Statistical Physics (Gostekhizdat, Moscow—Leningrad, 1951).
6. S. Lovell, Effect of Temperature on the Phase Composition and Strength of Frozen Ground [Russian translation]. Translation No. 12,657/9, All-Union Institute of Technical and Scientific Information of the State Scientific-Technical Committee (Sov. Min. SSSR and Akad. Nauk SSSR, Moscow, 1959).
7. Z. A. Nersesova, "Variation in the Iciness of Ground with Temperature, Dokl. Akad. Nauk SSSR 75(6) (1950).
8. Z. A. Nersesova, "Thawing of Ice in Ground at Negative Temperatures," Dokl. Akad. Nauk SSSR, 79(3):(1951).
9. Z. A. Nersesova, "Phase Composition of Water in Ground on Freezing and Thawing," Reports of the Frozen-Soil Research Laboratory," Collection I (Izd. Akad. Nauk SSSR, 1953).
10. S. G. Parkhomenko, Geocryology as the Study of Cryophilic Mountain Rocks. Transactions of the Committee on the Study of the Permafrost, Vol. VI [in Russian] (Izd. Akad. Nauk SSSR, Moscow—Leningrad, 1938).
11. J. K. Roberts, Heat and Thermodynamics [Russian translation] (Gostekhizdat, Moscow—Leningrad, 1950).
12. M. I. Sumgin et al., General Geocryology (Izd. Akad. Nauk SSSR, Moscow—Leningrad, 1940).
13. I. A. Tyutyunov, Variations and Transformations of Soils and Mountain Rocks at Negative Temperatures (Izd. Akad. Nauk SSSR, 1960).

14.   N. A. Tsytovich, Theory of the Equilibrium State of Water in Frozen Ground (Izd. Akad. Nauk SSSR, Ser. Geogr.) 9(5-6):(1945).

15.   P. A. Shumskii, Bases of Structural Ice Physics (Petrography of Fresh-Water Ice as a Method of Glaciological Study) (Izd. Akad. Nauk SSSR, 1955).

16.   P. S. Epshtein, Course of Thermodynamics (Gostekhizdat, Moscow—Leningrad, 1948).

PART TWO

EFFECT OF EXTERNAL ACTIONS ON THE PROCESSES OF CRYSTALLIZATION

NEW EXPERIMENTAL RESULTS ON THE ETCHING OF SINGLE CRYSTALS IN AN ULTRASONIC FIELD

A. P. Kapustin

The mechanism of the formation of etch figures is connected with various defects in the surface of the crystal lattice (including dislocations) of an actual crystal. The form of the etch figures is due to the different dissolution rates of crystals in different directions. The reasons for the greater vulnerability of individual sites on a crystal surface (etch figures) are now fairly well known. As a rule, there is a connection between the figures developing on the surface of a single crystal on etching and the outcrops of dislocation lines on this surface. In an actual crystal, the outcrops of dislocations constitute defective sites in which elementary pits are formed. The development of these leads to the appearance of etch figures. The etch method may thus be used together with other methods (optical, X-ray) to elucidate the structural perfection of crystals.

Experiment shows that the ordinary method of etching does not give the necessary resolving power, and requires a long time for the etch figures to appear.

The essence of the new method lies in the use of ultrasound, which affects the process of etch-figure formation [1-3]. Recently this method has been used to study the distribution of dislocations and residual stresses in semiconducting crystals [4].

It is known that, when ultrasound is passed into a solid situated in a liquid, action arises from the acoustic pressure varying periodically in the wave, from radiation pressure, and from artificial flows made possible by the "sonic wind." It is also known that, under certain conditions, the propagation of a sound wave is accompanied by cavitation phenomena. For a sufficiently large amount of ultrasound energy, the number of cavitation centers in the volume and at the solid—etchant boundary may be extremely large, and then cavitation takes on.

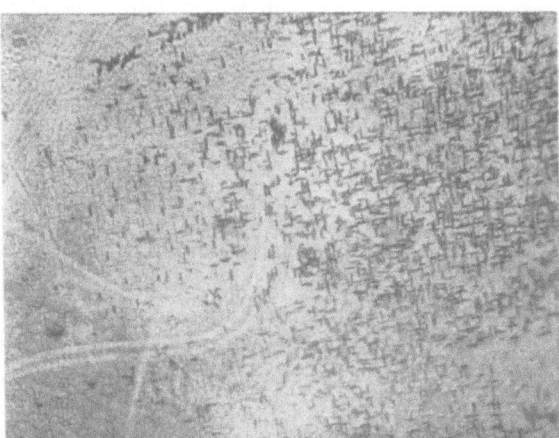

Fig. 1. Surface of a LiF single crystal: right) etch figures formed earlier at a pressure antinode; left) at a pressure node, where the effect of the ultrasound on their formation is weak.

135

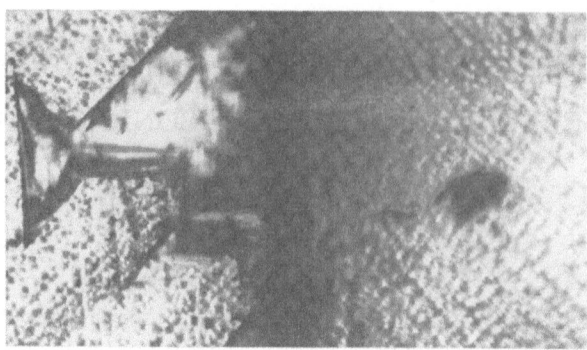

Fig. 2. Chains on etch figures on the two sides of a LiF
crystal split along the cleavage plane.

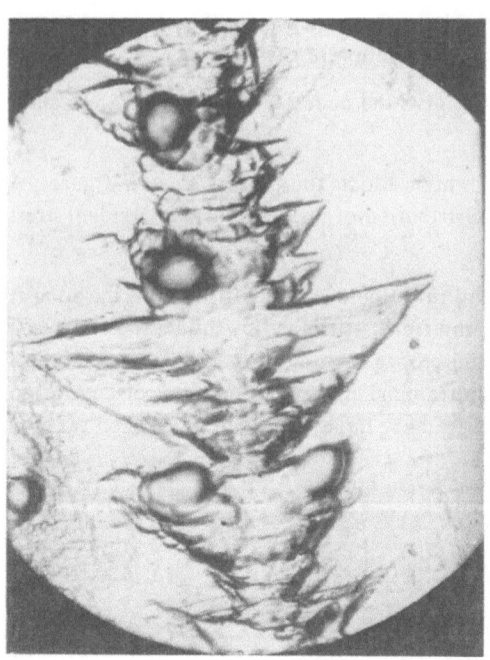

Fig. 3. Surface of plastically deformed LiF crystal. The pits on the edges of the scratch are revealed by the action of ultrasound.

an especially active character. The forces which operate clean the surface of the crystal from contamination, produce an intense circulation of the etchant, facilitate the removal of dissolution products, and so forth. Consequently, the etch figures develop in a shorter time and at any stage of dissolution of the crystal. It should also be noted that vibrations arise in the crystal itself under the influence of the ultrasound, and these may be very considerable if the characteristic frequency of the plate is in tune with the generator frequency. The existence of such oscillations may be judged from the nodal-line pattern [5].

We shall present some new results on the study of cavitation phenomena on single-crystal surfaces in an ultrasonic field. The ultrasound frequency was 20 and and 720 kc/sec; the ultrasound intensity was approximately 0.5 W/cm^2. The temperature of the liquid was 20 to 22°C. Since the two frequencies led to qualitatively identical results, we shall subsequently present data for the 20 kc/sec case only.

Crystals of lithium fluoride, potassium bichromate, terpin hydrate, germanium, etc., were submitted to the action of ultrasound. The etch figures of LiF have been thoroughly studied by the selective-etching method, and this enables us to compare known results with those given by ultrasonic irradiation.

The technique of preparing the crystals for irradiation and carrying out the actual experiments has been described earlier [3]. First of all, the long wavelength of the ultrasound enabled us to elucidate the action of its individual parts (nodes, antinodes) on the formation of etch figures. A single-crystal plate of LiF (50 × 20 × 1 mm) was set up perpendicular to the wave front so that distinct parts of it lay at a node and an antinode of the standing wave in the water. After irradiation the plate was removed, dried, and photographed in transmitted light. It transpired that the formation and development of the etch figures took place at the pressure antinodes of the standing wave. At the nodes, the ultrasound had practically no effect (Fig. 1). The increase in the density of the etch figures at the antinodal points can be explained only by the action of the ultrasonic field. By placing

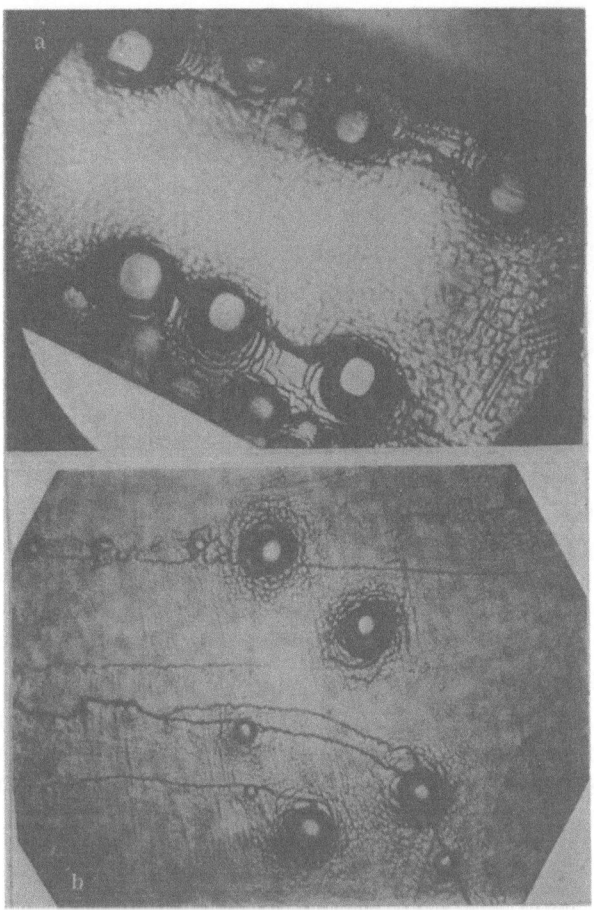

Fig. 4. a) Pits near the edges of the crystal plate; b) pits
formed at individual points of steps.

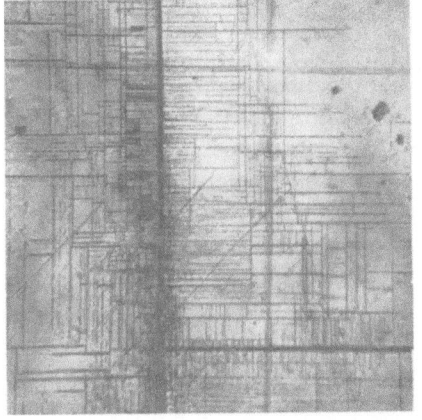

Fig. 5. Surface of LiF single crystal revealed by ultrasound after the action
of an electric field.

the crystal plate at the pressure antinode of the wave, the well-
known picture of edge and screw dislocations can be obtained.
It should be noted that there is always a large number of etch
figures on the crystal faces subjected to the action of ultrasound
and deformed in the course of etching, and moreover they appear
on these faces much more quickly.

Let us see how etch figures are disposed on the planes of a
cleaved crystal. For this the two parts of a crystal split along the
cleavage plane were simultaneously irradiated by ultrasound. The
irradiation revealed oriented networks on both sides of the cleavage, and these coincided on putting the plates together (Fig. 2).
We attribute the origin of these to plastic strain induced in the
crystal on cleavage. The agreement between the chains of etch
figures (tracks) on the two sides of the cleavage indicates that
the etch figures were formed at dislocations.

It is interesting to study the etch figures on the faces of
plastically strained crystals. In this case we may expect distortion of the etch figures. To perform the experiment, a scratch

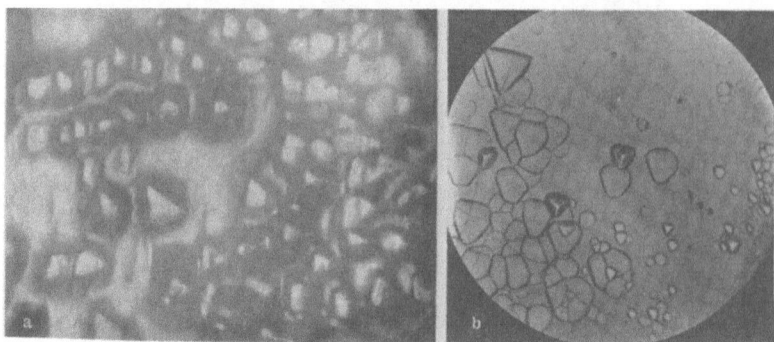

Fig. 6. Etch figures of a) potassium bichromate; b) germanium.

was implanted on a perfect crystal face with a needle. Under the microscope this appeared as a chain of tri-angular figures of different sizes, arranged vertex-to-base. Then the crystal was irradiated by ultrasound in water. The figures appearing at the scratch differed from those ordinarily formed on irradiating on unstrained crystal. Moreover, at roughly equal distances along the scratch, oval pits developed (Fig. 3). Similar pits appeared on crystals not deliberately deformed subjected to ultrasound, but only near the edges of the crystal plate (Fig.4a). In the central region they occurred more rarely, and then mainly at individual points of steps (Fig. 4b).

On the surface of a LiF crystal ultrasound also enables us to obtain an interesting picture resulting from the disruptive effect of an electric field. In order to produce this, a crystal plate ($15 \times 10 \times 1$ mm) was placed between electrodes in an inhomogeneous electric field created by a small induction coil. A discharge occurred for 10 to 12 sec. In this case, the structure revealed on the plate by the effect of ultrasound constituted a multitude of fine branches (Fig. 5), somewhat reminiscent of that in the work of Gilman and Strauff [6].

We used ultrasound to obtain microscopic etch figures of potassium bichromate, terpin monohydrate, and other crystals. Thus Fig. 6a shows etch figures of potassium bichromate revealed by ultrasonic etching at the face of a crystal immersed in glycerine. Then etch figures were obtained for terpin monohydrate by irradiating the crystal with ultrasound in water at a face perpendicular to the polar axis. The irradiation time was 4 to 5 min. Figure 6b shows microphotographs of germanium. The revelation of the etch figures in the ultrasonic field was effected on a single [111] face. The etchant used was a solution of $K_3[Fe(CH)_6]KOH$. The solution temperature was 22°C. The figures began to appear fairly rapidly and after 7 to 10 min covered the face of the crystal.

Our experiments showed that, apart from ordinary etch figures corresponding to dislocations, ultrasound also reveals other figures, apparently caused by some other kind of crystal-lattice defect. This may be useful in examining dislocation-free crystals. In all cases, the time needed to reveal the etch figures is considerably reduced, even though a dilute etchant may be used. Irradiation by ultrasound leads to more uniform action of the etchant on the surface of the crystal. It should be noted that ultrasound apparently not only reveals dislocations but also initiates their motion. It should thus be possible to develop a method which would permit the number and disposition of dislocations in crystals to be controlled, and this is of undoubted interest.

Literature Cited

1.   Kh. S. Bagdasarov and A. P. Kapustin, "The production of etch figures by means of ultrasonic oscillations," Kristallografiya 1(1):139-40 (1956).
2.   Kh. S. Bagdasarov and V. Ya. Khaimov-Mal'kov, "Some experimental data on the nature of etch-figure formation in an ultrasonic field," Kristallografiya 2(2):309-10 (1957).
3.   A. P. Kapustin, "The observation of dislocations by ultrasound," Kristallografiya 4(2):265-67 (1959).
4.   Z. Dragoun, Zvlǎstni vytisk ze sborniky 3(56):(1960), 215-27. Experimentǎlni studie vlivu ultravukoveno pole na leptǎní monokrystalickéno germania a kremiku.
5.   Bergman, Ultrasound and Its Use in Science and Technology [Russian translation] (IL, 1956).
6.   T. T. Gilman and D. W. Strauff, Nucleation of Dislocations Accompanying Electric Breakdown in LiF Crystals, J. Appl. Phys., Vol. 29(2):(1958).

DISPERSION HARDENING OF LEAD-BASE ALLOYS
IN AN ULTRASONIC FIELD

F. K. Gorskii and V. I. Efremov

The effect of ultrasound on the kinetics of dispersion hardening is a solidly established experimental fact [1-5].

Considering the decomposition of supersaturated solid solutions from the point of view of the general laws of phase transformations, the reason for the influence of ultrasound must be sought in the changes which it induces in the interphase energy at the boundary between a nucleus of the hardening phase and the alloy, i.e., the work of formation of the nucleus and the activation energy of the diffusion processes.

An increase in the numbers of nuclei of the new phase may also be associated with the formation of dislocations under the influence of ultrasound [7].

In a paper by Larikov and Polotskii [8] it was suggested that the acceleration of aging in aluminum alloys in an ultrasonic field was a secondary effect: The ultrasound did not directly influence the kinetics of the process, but produced local overheating at grain boundaries, the acceleration of decomposition being simply the result of increased temperature within the sample. In this investigation a power of 200 W at frequency 750 kc/sec was used. The first hardness measurements for Duralumin samples irradiated by ultrasound and heated to 74° in a thermostat were made 2 h after quenching. The attainment of like hardness, however, does not mean that the form of the curves in the interval 0 to 120 min was the same (see Fig. 4 in [8]). Larikov and Polotskii also found no effect of ultrasound on the dispersion hardening of a lead−6% tin alloy.

Starting from the principle that ultrasound acts on the decomposition of solid solutions by facilitating diffusion processes, the authors studied the dispersion hardening of very diverse lead−base alloys, using as an ultrasound source a magnetostriction emitter with a frequency of 19 to 20 kc/sec and power of the order of 1.25 kW, which produced a considerable increase in the amplitude of the ultrasonic oscillations. The choice of lead−tin alloys for examination was made because of the low sensitivity of aging kinetics to temperature, since at room temperature there is a flat maximum in the solid-solution decomposition rate for the 6% tin alloy.

The alloys were prepared from reagents of "pure" standard [6]. Samples 70 × 50 × 8 mm in size were cast simultaneously in tens in a cast-iron chill mold. The surface of the cast sample did not require further treatment; only the sprue was removed. The hardness measurements were made by the Brineil method on a mechanical press of the TSh type with a 10 mm diameter sphere loaded with 250 kg for 10 sec (HB 10/250/10). The control sample was kept directly in the air; the sample subjected to the ultrasound was situated in a bath with running water. The water-flow rate was sufficient for the rapid removal of the heat evolved in the sample by the ultrasound. The ultrasound was transmitted to the sample bath from a magnetostriction converter of the PMS-6 type with a square 300 × 300 mm diaphragm.

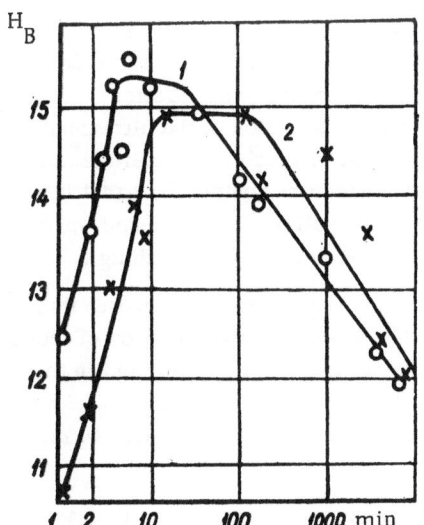

Fig. 1. Hardness variation in aging 90% Pb−10% Sn alloys: 1) in ultrasound; 2) natural aging.

139

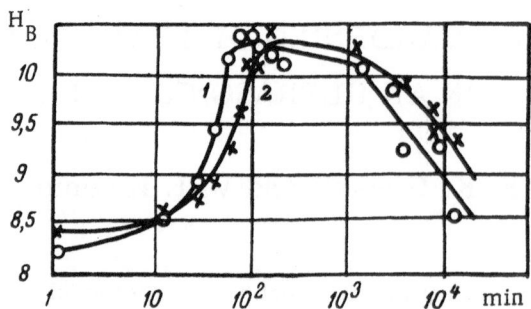

Fig. 2. Hardness variation in the aging of 94% Pb−6% Sn alloys: 1) in ultrasound; 2) natural aging.

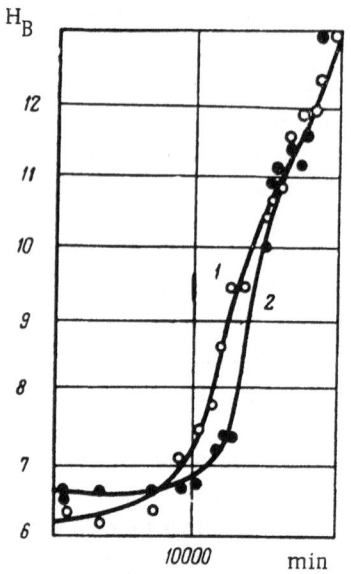

Fig. 3. Hardness variation in aging of a 98.5% Pb−1.5% Sb alloy: 1) in the ultrasound; 2) natural aging.

The converter was fed from an industrial ultrasonic generator UZG-2.5A at a frequency of 19 to 20 kc/sec and an acoustic output to the bath of 1.25 kW.

According to the phase diagram, the maximum solubility of tin in lead is 19.5%. We prepared alloys with 15, 10, and 6% tin. The samples were water-quenched after a 6-h homogenization from 180°C. The alloy containing 15% tin gave such a rapid decomposition of the solid solution that at the first moment of observation only a fall in hardness, associated with coagulation of the particles, was observed; this was therefore not studied in detail, although mention should be made of the more rapid fall in the hardness of the samples subjected to the ultrasound. The acceleration of coagulation in the ultrasonic field is a result of the easing of the diffusion processes.

The decomposition of the solid solution takes place in three stages. In the first stage we have chiefly the nucleation of the new phase, and the concentration changes very little. In the second change we have mainly the growth of the nuclei generated, and the supersaturation falls sharply. In the third phase the concentrations finally level out. It is natural to suppose that the action of the ultrasound appears in the first phase, namely, that in which the nuclei are formed. Hence in our experiments the ultrasound was applied for 3 to 4 h from the moment of quenching (for antimony alloys the ultrasound was applied for 2 to 4 h per day for the first few days after quenching).

The results of the experiments with the 10%−tin alloys appear in Fig. 1. The kinetic curve passes through a maximum even for the samples subjected to the ultrasound, being shifted as a whole in the direction of shorter times, as measured from the start of the process. In the control sample, the aging ends in 10 min, and in the sample subjected to ultrasound in 4 min, i.e., the ultrasound accelerates the aging process by a factor of 2.5. Analogously, the sample subjected to ultrasound passes into the overaging state (falling hardness) after 40 min, whereas the control sample does this after 150 min (the acceleration factor is 3.7).

The alloy containing 6% tin was studied in more detail. Preliminary experiments established that at 50°C the decomposition of the solid solution was roughly half as fast as aging at room temperature, while 100°C there was zero supersaturation and no decomposition. Thus any amount of local superheating would lead to slower rather than faster decomposition. In Fig. 2, however, we see that in the case of the 6% alloy ultrasound accelerates both aging (control in 150 min, under ultrasound 70 min) and overaging (control after 2000 min, under ultrasound 1000 min) by a factor of two. Both processes — the precipitation of the hardening phase

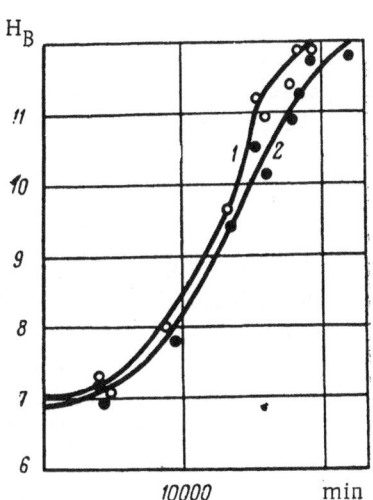

Fig. 4. Hardness variation in aging
a 98% Pb−2% Sb alloy: 1) in ultrasound; 2) natural aging.

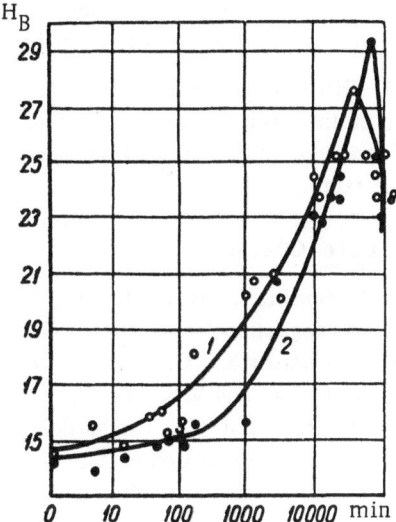

Fig. 5. Variation in hardness in aging
calcium Babbitt (1% Ca, 0.8% Na, balance Pb): 1) in ultrasound; 2) natural aging.

on aging and its coagulation on overaging — are accelerated by ultrasound. Figures 3 and 4 show the aging of lead alloys containing 1.5 and 2% Sb respectively. Before quenching, the alloys were given a 6-h homogen-ization at 245°C. The alloy with 1.5% antimony came to its maximum hardness in 61 days, while the ultra-sound accelerated the decomposition by a factor of 1.7. The alloy with 2% antimony attained maximum hard-ness in 57 days, and ultrasound accelerated the decomposition by a factor of 1.8. Owing to the slowness of the processes, the examination of these alloys was not extended to overaging.

For calcium Babbitt metal (1% Ca, 0.8% Na, balance Pb) the results appear in Fig. 5. Before quenching, homogenization at 190° was carried out for 6 h. In contrast to the antimony alloys, the calcium Babbitt came to maximum hardness far earlier, and showed sharp overaging. Ultrasound accelerated all the processes in this case by a factor of 1.7.

It follows from the experiments that ultrasound accelerates the dispersion hardening of lead alloys of very different compositions, as in the case of Duralumin.

The fact there is really an acceleration in the aging of the alloys studied contradicts the concept of Polotskii and Larikov regarding the thermal nature of the process. The different sensitivities of alloys to the effects exerted by various physical factors on the kinetics of their solid-solution decomposition constitute in-dividual properties of the alloys. Apparently alloys which are sensitive to certain factors are sensitive to other factors to the same degree. Thus, for example, Hilliard and Cahn [9] showed that pressures of 30,000 atm lowered the rate of Θ-phase precipitation in Duralumin at at 400°C by a factor of 40, but the same pressure has neg-ligible effect on the aging of lead−tin and copper−nickel alloys.

Conclusions

1. Leading causes underlying the accelerated decomposition of solid solutions under the action of ultra-sound include the reduction in the work of forming nuclei of the hardening phase, and the consequent rise in the number of nuclei, as well as the easing of diffusion processes (reduction of the activation energy). The evolution of heat in the samples can only play a minor role.

2. The principle that the effect of ultrasound on the kinetics of solid-solution decomposition is simply a matter of increasing the number of nuclei of the new phases means that ultrasound is only needed in the initial

stages of the process, when the generation of new-phase centers is the leading factor. In subsequent stages there is no need to apply ultrasound in order to speed the process.

3. Diffusion processes are affected by temperature and pressure as well as by ultrasound. Hence the effects of these on the aging process are related. When aging depends on temperature, it will also depend on the other factors (pressure and ultrasound), as indeed occurs for aluminum alloys. Lead-base alloys are less sensitive to all these factors.

Literature Cited

1.   F K. Gorskii and G. P. Efremov, Izv. Akad. Nauk Belorus. SSR (3):155-64 (1953).
2.   K. M. Pogodina-Alekseeva and G. P. Éskin, Metallovedenie i Obrabotka Metallov (1):42-45 (1956).
3.   N. T. Gudtsov and M. N. Gavze, Zh. Neorgan. Khimi. (7):1533 (1956).
4.   É. A. Al'ftan and V. S. Ermakov, Akust. Zhurn. (4):307-14 (1958).
5.   K. V. Gorev and P. A. Parkhutik, Third Annual Scientific-Technical Conference. Application of Ultrasound in the Production of Alloys and Their Heat Treatment, No. 2 (Moscow, 1962), pp. 21-28.
6.   F. K. Gorskii and V. I. Efremov, Ibid., pp. 28-33.
7.   T. Kh. Chormonov, Ibid., pp. 3-7.
8.   L. N. Larikov and P. G. Polotskii, Collection of Scientific Works, No. 9, Academy of Sciences, Ukrainian SSR, Institute of Metallophysics, Vopr. Metalloved. i Fiz. Met., pp. 50-53 (1960).
9.   S. E. Hilliard and T. W. Cahn, Progress with Very High Pressure Resistivity (New York—London, 1961), pp. 109-124.

ROLE OF INSOLUBLE IMPURITIES IN THE CRYSTALLIZATION OF METALS IN AN ULTRASONIC FIELD

O. V. Abramov and I. I. Teumin

The special role of impurities in facilitating the refinement of the cast grain of a metal in a field of ultrasonic oscillations (u.o.) was indicated in [1-3].

In this article we shall study the effect of insoluble impurities on reducing the value of threshold power P_n in treating pure metals (tin, bismuth, zinc, aluminum), and shall consider some possible mechanisms of this effect.

Weighed portions of 70 to 100 g metal were melted in alundum crucibles. The metal was superheated to the following temperatures: tin 250°, bismuth 310°, zinc 460°, aluminum 680°C. The superheating temperature was selected in such a way as to make the control ingots coarse-grained in macrostructure.

The crucible containing the superheated melt was set on a water-cooled copper stand, ensuring directional solidification from the bottom of the crucible.

The elastic oscillations were introduced into the melt from above. The u.o. source was a magnetostriction converter fed from an ultrasonic 10 kW generator. The power introduced into the melt was measured by the method described in [4].

As insoluble impurities we selected: for tin SiO_2, for bismuth Al_2O_3, for zinc SiO_2, and for aluminum W. The dimensions of the impurity particles were on the order of 0.05 mm. Introduction of these impurities in 0.5% quantities into the melts in the absence of u.o. did not cause any changes in the structure of the ingot as compared with the control.

For effective introduction of the impurity and even distribution over the volume of the ingot, the melt containing the impurity was treated with u.o. above the crystallization temperature. In subsequent experiments, the ingot containing the impurity so introduced was again melted and superheated to the necessary temperature, after which the metal was either crystallized under ordinary conditions or else subjected to elastic oscillations in the course of crystallization.

Measurements showed that the introduction of insoluble impurity reduced the value of the threshold power (Table 1).

TABLE 1. Effect of Insoluble Impurities on the Value
of Ultrasound Threshold Power

Metal	Threshold power		Impurity
	without impurity	with impurity	
Sn	60	45	SiO_2
Bi	30	20	Al_2O_3
Zn	160	100	SiO_2
Al	70	50	W

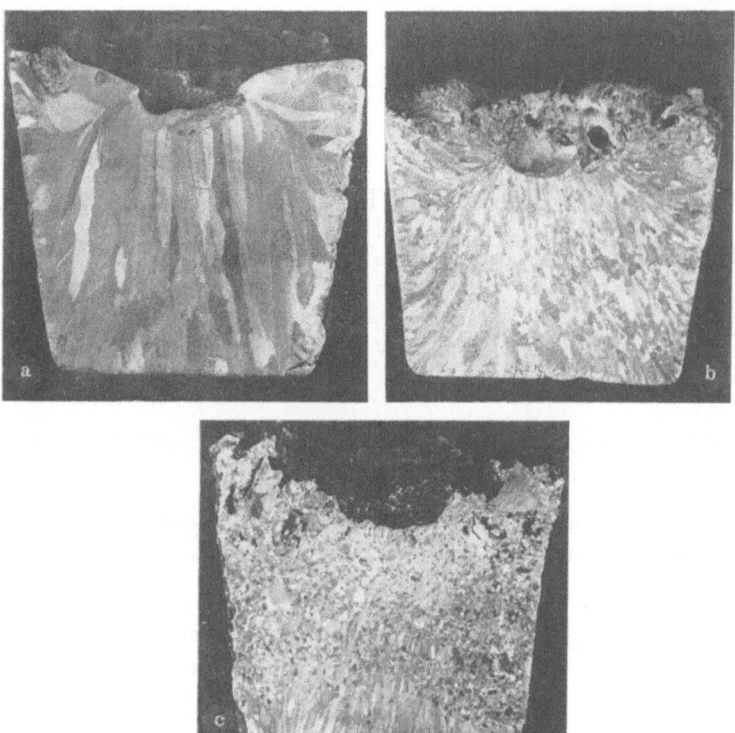

Fig. 1. Structure of zinc ingots (× 2): a) control sample; 2) subjected to
160 W u.o.; c) subjected to 100 W u.o. with introduction of 0.5% SiO$_2$.

Photographs of the control zinc specimen and those subjected to threshold power with and without impurity appear in Fig. 1a, b, and c.

It should be mentioned that the refinement in the structure of the impurity-containing metal subjected to u.o. was considerably greater than that of the pure metal subjected to the same or greater power. Thus u.o. treatment of pure tin at 120 W refined the grain by a factor of approximately 20, while for tin containing 0.5% SiO$_2$ a power of 90 W reduced grain size by a factor of only 50 to 70.

In considering possible causes for the effect of insoluble impurities on the fall in the threshold power, a number of suggestions may be made.

1. Cavitation phenomena, which to some extent are responsible for the change in the structure of the metal, begin to arise at lower u.o. powers in the melt on introducing impurities (owing to the disruption of homogeneity in the melt).

2. In the u.o. field, the impurities introduced into the melt become "activated," i.e., impurity particles are transformed into crystallization centers by some "specific" action of the u.o.

It is known that cavitation phenomena arise in a liquid if the pressure therein (sum of the hydrostatic and sonic pressures) at some moment of time exceed the cavitation threshold in absolute magnitude. The cavitation threshold of a liquid, in particular, depends on the liquid—vapor surface tension.

The relationship between the cavitation threshold (pressure at which the liquid breaks up with the formation of a cavitation vacancy) and the surface tension and temperature of the liquid is derived theoretically in [5].

Fig. 2. Structure of a tin ingot with 0.5% impurity after remelting with superheating of 1 to 2°C (× 2): a) ingot subjected to u.o.; b) control ingot.

If a cavity develops not inside the liquid but as a result of its breaking off from the surface of a solid (in particular, from an impurity particle), the value of the negative pressure needed to form the cavity changes substantially. For complete nonwetting, the resistance of the liquid to breaking off from the solid tends to zero. Thus the introduction of impurity into the melt may considerably reduce the cavitation threshold.

By "activation" of the impurity we mean such a change in the state of its surface under the influence of the ultrasonic field in the metal melt that the impurity particle is able to become a crystallization center.

It was shown in [6] that in a u.o. field solids may be metallized by a liquid which does not effect such metallization under ordinary conditions.

In order to check the idea that impurity particles were metallized in an ultrasonic field, we decided to examine tin containing Al_2O_3, SiO_2, and W impurities, which have no effect on the crystal structure of tin in the absence of u.o.

First, experiments were made on the metallization of plates of these materials by tin in an ultrasonic field. The plates, of area $S = 0.5$ cm^2, were placed in the melt at a distance of 10 mm from the end of the emitter. U.o. of maximum intensity were introduced into the melt. The treatment took place at 240°C for 3 min. Regions metallized by the tin were observed on the plates, firmly connected to the surface. Holding these plates in the melt in the absence of the ultrasonic field caused no metallization to ensue.

Introducing 0.5% impurities of these substances, with particle size not exceeding 0.05 mm, into a melt treated with u.o. led to a reduction in the value of the threshold power.

Thus we may suppose that the particles are metallized in an ultrasonic field and converted into crystallization centers. For an indirect confirmation of our conclusions on the activation of impurities, we performed experiments on their deactivation by superheating the melt. Superheating the metal treated with u.o. and containing activated Al_2O_3 impurity by 1 to 2° over the crystallization temperature did not lead to deactivation of the impurities; the ingot structure remained fine (Fig. 2a). The same superheating of untreated metal containing impurity gave a relatively coarse grain (Fig. 2b).

Superheating the u.o.-treated metal containing activated impurity by 5 to 7° over the crystallization temperature led to the formation of a coarse grain. Treating the melt with 45 W u.o. (impurity added) at 1 to 2° over the crystallization temperature led to a refinement of the structure, but the degree of this refinement was smaller than when the melt was treated up to complete solidification of the ingot.

These experiments confirm that metallization of impurity particles, followed by their conversion into crystallization centers, occurs in an ultrasonic field during the process of crystallization. The metallized layer formed, now situated in molecular contact with the impurity, has a melting point higher than the main metal [9]. Superheating by several degrees is nevertheless sufficient for the deactivation of the impurity activated in the ultrasonic field.

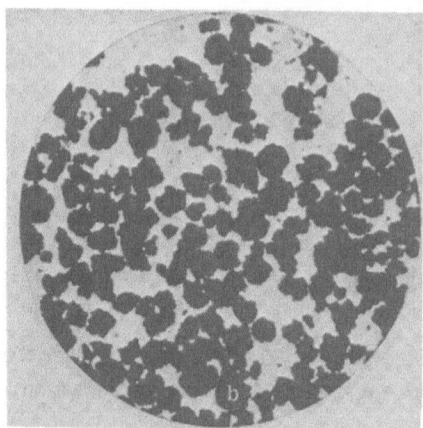
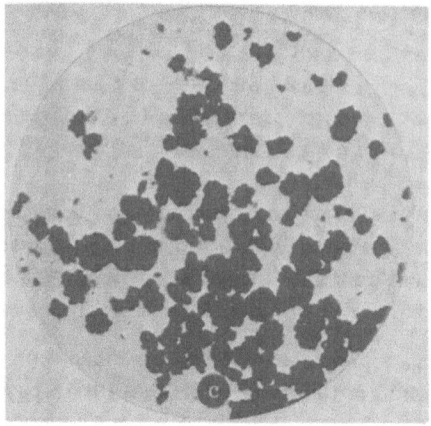

Fig. 3. Scheme for ascertaining the dispersion of impurities in an ultrasonic field: a) structure of aluminum containing 0.5% Al_2O_3 treated with maximum power u.o. ($\times$ 2); b) particles of Al_2O_3 separated out from the ingot ($\times$ 180); c) particles of Al_2O_3 before introduction into the melt ($\times$ 180).

The mechanism of the metallization is not yet quite clear, but we may make a few suggestions:

a. U.o. cleans the surface of the solid from possible scab and contaminations, and the intense diffusion processes at the solid/liquid interface taking place in the ultrasonic field facilitate the formation of a layer wetted by the liquid.

b. It is known that ordinary methods of breaking up a solid give particles with a severely distorted boundary layer, while the dispersion of a solid in an ultrasonic field gives particles with very little structural disruption [7]. It is possible that, on the introduction of impurities into a melt being subjected to u.o., the dispersive action removes the heavily distorted layer from the impurity particles. In that case the impurity can become a crystallization center to the extent determined by its isomorphism with the crystallizing metal.

c. Still another possible impurity-activation mechanism is indicated in [8]; this is based on the effect of viscous friction between the surface of the impurity particles and the melt in the ultrasonic field. The tangential forces of viscous friction may change the state of the transitional layer and reduce the surface tension at the impurity/melt boundary. It follows from analysis of impurity-particle motion in a liquid in an ultrasonic field that the viscous frictional forces will be the greater, the more the oscillatory velocity of the particle differs from the velocity of the liquid.

Theoretical consideration of the motion of a solid particle in a liquid subjected to ultrasonic treatment leads to König's formula [8]:

$$\dot{\xi}_s = \frac{1 + \sqrt{a} + j\sqrt{a}\left(1 + \frac{2}{3}\sqrt{a}\right)}{1 + \sqrt{a} + j\sqrt{a}\,(1 + b\sqrt{a})}\,\dot{\xi}_{l} \cdots$$

$$(1)$$

TABLE 2. Effect of Impurities on the Structure of Aluminum and Tin

Melt	Impurity	$\dfrac{\rho_s}{\rho}$	$\dfrac{\breve{\xi}_s}{\breve{\xi}_1}$	$\dfrac{a_c}{a_{tr}}$
Sn	B W	0.33 2.76	2.0 0.5	6 7
Al	B W	1.00 8.00	1.0 0 2	1 5

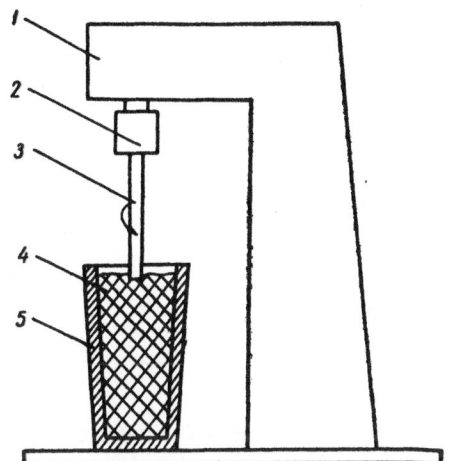

Fig. 4. Apparatus for solidification of metals in a field of turbulent flows: 1) drilling machine; 2) chuck; 3) rotating rod; 4) solidifying melt; 5) mold.

where $\breve{\xi}_s$ is the oscillatory velocity of the solid-phase particle, $\breve{\xi}_1$ is the oscillatory velocity of the liquid phase, $a = \omega \rho r^2/2\eta$; $b = (2/9)(1+2\ \rho_s/\rho)$, r = particle radius, ω = oscillation frequency, ρ = density of the liquid phase, ρ_s = density of the solid phase, and η = viscosity of the liquid.

A more complete analysis of the question [8], made with an allowance for the elastic properties of the solid and liquid phases, leads to a more precise expression for ξ_s. We shall restrict our discussion, however, to the formula given.

Analysis of the formula shows that, the more strongly the densities of the solid and liquid phases differ, the greater will be the difference in their oscillatory velocities, and the greater will be the effect introduced by the forces of viscous friction.

In order to check the possible mechanism of activation by removing the distorted layer from the impurity particles in an ultrasonic field, we selected as impurity TiO_2, which has a certain isomorphism with tin but does not affect its structure in the absence of u.o.

The introduction of 0.5% of this impurity during treatment of the crystallizing tin with 45 W u.o. led to a considerable refinement of the structure. Subjecting the tin melt containing the impurity to u.o. of maximum power at 250°, when a fair amount of mechanical action might be expected on the surface of the impurity particles, and subsequently solidifying the ingot in the absence of ultrasound, produced no change in structure as compared with the control. We thus see that the idea of the distorted layer being removed from the impurity particles in an ultrasonic field cannot be justified experimentally.

Direct proof of these conclusions was made with aluminum, using 0.5% Al_2O_3 as impurity. The structure of an Al ingot containing 0.5% Al_2O_3 and treated with maximum power is shown in Fig. 3a.

After treatment of the metal with u.o., the impurity was separated out by dissolving the ingot in 15% NaOH solution. The separated impurity was washed in 10% HCl and analyzed under the microscope. No difference was found between the sizes and shapes of the impurity particles separated out from the u.o.-treated ingot and particles which had not been introduced into the melt (Fig. 3b and c).

In order to substantiate the proposition regarding the role of viscous frictional forces, we selected impurities with densities differently related to those of the melts. Experiments were made on tin and aluminum,

Fig. 5. Structure of aluminum AVOOO ingots solidified
with a rotating rod in the melt: a) without impurity; b)
with 0.5% W added.

the impurities being tungsten and boron, which have no effect on the structure of the metals in the absence of
u.o. The introduction of these impurities into the u.o.-treated melt led to a refinement in the structure of tin
on adding 0.5% W and B, and to a refinement of the cast structure of aluminum with the addition of tungsten.

From the data thus obtained we determined the degree of grain-size reduction a_c/a_{tr} (a_c = mean grain
size of control sample, a_{tr} = that of the treated sample). For these we used formula (1) to calculate the ratio
of the amplitudes of the oscillatory velocities ξ_s and ξ_1 of the solid and liquid phases. These velocities, in
particular, are determined by the ratios of the phase densities ρ_s and ρ. Table 2 shows results of these measurements and calculations. It follows from the table that, as the ratio ρ_s/ρ departs from unity, the role of the
viscous forces increases, and so does the refinement of the grain.

In order to confirm the data on the effect of viscous friction, and also to elucidate the importance of
its role in obtaining fine crystal structure in a u.o. field, we chose crystallization conditions under which all
factors characteristic of the ultrasonic effect and determining a dispersion mechanism unconnected with viscous
friction were excluded. Such factors, as we know, include: a) cavitation processes; b) oscillatory pressures at
the solid/liquid boundary; c) gradients of oscillatory pressures.

Such conditions may be achieved, for example, by creating turbulent flows in the melt (Fig. 4).

A 10 mm diameter rod rotating at 4200 rpm around its axis was introduced into 200 cm^3 of molten metal
poured into a mold. In order to eliminate the possibility of metal solidifying on the rod, the latter was previously heated to a temperature above the crystallization point of the metal.

Experiments were made with aluminum AVOOO. Despite the rotatory treatment, the structure of the
metal remained coarse—crystalline (Fig. 5a). After the introduction of 0.5% W into the melt and the application of rotatory treatment, fine-grained structure developed (Fig. 5b).

In the control ingot, the structure remained coarse-grained. Thus in experiments with turbulent flows
we obtained the same effect of impurity activation as in the u.o. field.

These experiments enable us to conclude that the activation of impurities in an ultrasonic field occurs
by way of metallization of the particles under the influence of viscous frictional forces.

Conclusions

1. The value of the threshold power P_n of elastic oscillations is considerably reduced on introducing insoluble impurities into the treated melt, and the degree of grain refinement exceeds that obtained by u.o. treatment in the absence of impurities.

2. The effect of elastic oscillations on the insoluble impurities in the course of crystallization is to activate these impurities.

3. Some possible activation mechanisms have been considered, and we have shown the importance of the forces of viscous friction in metallizing the impurity particles present in the melt subjected to the ultrasonic field.

4. We have considered other possible effects of u.o. on the impurities. We have shown that such factors as the dispersion of the impurity particles and changes in their shape do not occur. Effects associated with the cavitation process are not necessary for activation.

Literature Cited

1. V. I. Danilov and G. Kh. Chedzhemov, Probl. Metalloved. i Fiz. Met. (4):34-49 (1955).
2. N. N. Sirota, E. A. Lekhtblau, and É. M. Smolyarenko, Fiz. Met. i Metalloved. 7(6):879-83 (1959).
3. O. V. Abramov, V. E. Neimark, and I. I. Teumin, Fiz. Met. i Metalloved. 13(6):(1962).
4. O. V. Abramov and I. I. Teumin, Collection of Contributions to the Second All-Union Conference on the Use of Ultrasonics in the Production and Heat Treatment of Alloys (NTO, Mashprom, 1962).
5. J. Fisher, J. Appl. Phys. 19(11):1062 (1948).
6. G. I. Pogodin-Alekseev, Ultrasound and Low-Frequency Vibrations in the Production of Alloys (NTO, Mashprom, 1961), p. 25.
7. A. Crawford, Ultrasonic Techniques [Russian translation] (IL, 1958), p. 193.
8. I. I. Teumin, Probl. Metalloved. i Fiz. Met. (7):375-416 (1962).
9. V. E. Neimark, Rare-Earth Elements in Steels and Alloys (Metallurgizdat, 1959), p. 231.

KINETICS OF THE DECOMPOSITION OF SUPERSATURATED SOLUTIONS OF ALUMINUM FLUORIDE IN ULTRASONIC FIELDS

Yu. N. Tyurin and S. I. Rempel'

Aqueous supersaturated solutions of aluminum fluoride formed as a result of the neutralization of 10% H_2SiF_6 by aluminum hydroxide in the reaction

$$H_2SiF_6 + 2Al(OH)_3 \longrightarrow SiO_2 + 2AlF_3 + 4H_2O$$

have great stability. Thus the decomposition time of the supersaturated solution, resulting in the precipitation of crystalline $AlF_3 \cdot 3.5H_2O$ is 10 to 14 h at 90°. Under the influence of ultrasound (frequency 25 kc/sec), the stability of the supersaturated solutions of aluminum fluoride decreases sharply, and the decomposition time diminishes to 2 to 4 h, depending on the intensity of the oscillations [1]. Experiments carried out by the authors showed that low-frequency ultrasound causes dispersion of the $AlF_3 \cdot 3.5H_2O$ crystals being formed, and hence an increase in the number of crystallization centers.

Under the influence of 800 kc/sec ultrasound, the stability of the supersaturated solution also falls, but to a smaller extent; instead of dispersion, there occurs a coagulation of the forming nuclei, so that the aluminum fluoride develops as spherulitic aggregates.

At the present time there is quite a wide range of experimental material on the effect of ultrasound in crystallization processes [2], but the kinetics of the decomposition of supersaturated solutions in ultrasonic fields, constituting a particular case of mass crystallization, has been little studied.

Let us check the applicability of the kinetic equations derived for other cases of crystallization in aqueous solutions to a description of the kinetics of the decomposition of supersaturated aluminum fluoride solutions in ultrasonic fields.

If the ultrasound does not have any specific effect on the crystallization and its influence resides in the acceleration of processes accompanying the formation and growth of crystals, then the kinetics of the decomposition of supersaturated solutions should be described by the kinetic equations derived for cases of crystallization without ultrasound. One of these equations, suitable for describing the decomposition process of aluminum fluoride solutions, was derived by Roginskii and Todes [3], who demonstrated theoretically that the crystallization of aqueous solutions of salts without seeds obeyed the relationship

$$At = \ln \frac{\sqrt{1 + \sqrt[3]{1 - \frac{x}{x_n}} + \sqrt[3]{\left(1 - \frac{x}{x_n}\right)^2}}}{1 - \sqrt[3]{1 - \frac{x}{x_n}}} + \sqrt{3}\left(\operatorname{arctg} \frac{\sqrt[3]{1 - \frac{x}{x_n}} + 1/2}{\sqrt{3/4}} - \frac{\pi}{6}\right),$$

$$A = 3\sqrt[3]{\gamma \rho N}\, K X^{2/3},$$

(1)

where x/x_N is the relative supersaturation, γ is the form factor of the crystal, K is a constant connected with the diffusion coefficient, and ρ is the density of the crystals.

151

TABLE 1. Effect of Ultrasound on Relative Supersaturation of Aluminum Fluoride Solution

Expt. No. 1 without ultrasound, $C_0 = 1.05$; $X_0 = 0.212$		Expt. No. 2, ultrasound, 800 kc/sec, intensity 1 W/cm^2, $C_0 = 1.05$; $X_0 = 0.280$		Expt. No. 3, ultrasound, 800 kc/sec, intensity 3 W/cm^2, $C_0 = 1.04$; $X_0 = 0.290$		Expt. No. 4, ultrasound, 25 kc/sec, intensity 1 W/cm^2, $C_0 = 1.26$; $X_0 = 0.280$	
Time from start of expt., min	Supersaturation X, moles/liter	Time from start of expt., min	Supersaturation X, moles/liter	Time from start of expt., min	Supersaturation X, moles/liter	Time from start of expt., min	Supersaturation X, moles/liter
0	0.838	0	0.770	0	0.750	0	0.980
120	0.783	90	0.690	60	0.690	60	0.880
150	0.770	150	0.483	90	0.658	90	0.850
240	0.610	180	0.313	120	0.555	120	0.770
300	0.561	240	0.172	150	0.372	150	0.565
360	0.448	360	0.137	180	0.246	180	0.244
480	0.230	480	0.090	210	0.210	210	—
540	0.210	540	0.053	270	0.080	240	0.160
630	0.128	—	—	—	—	300	0.114
660	0.068	—	—	—	—	360	0.068

Note. X_0 and C_0 are in moles/liter.

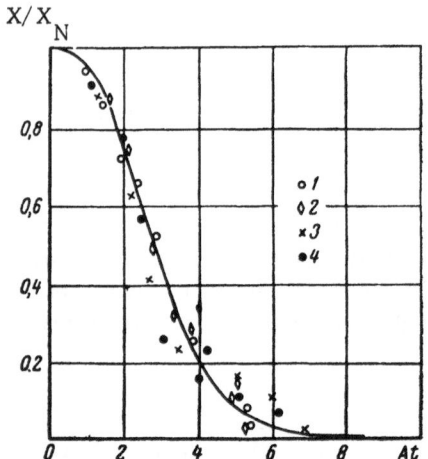

X/X_N

Variation of relative supersaturation with time: 1) expt. No. 1; 2) expt. No. 2; 3) expt. No. 3; 4) expt. No. 4.

This relation corresponds to an S-shaped curve. By means of the Roginskii—Todes equation the kinetics of the decomposition of supersaturated solutions may be described independently of the decomposition time, since the values of the dimensionless time, At, calculated theoretically enable us to compare them with those found experimentally by the equation

$$A = \frac{At_1}{t_{1\exp}} = \frac{At_2}{t_{2\exp}} = \ldots = \frac{At_n}{t_{n\exp}},$$

where $t_{1\exp}$, $t_{2\exp}$, $t_{n\exp}$ give the experimental time.

Table 1 shows the results of four experiments on the decomposition of stable supersaturated solutions of aluminum fluoride in ultrasonic fields for the same initial concentration C_0 and initial supersaturation X_0, calculated by the method described by Deryabina and Mishchenko [4]. The decomposition temperature in all experiments was 90°. Experiment 1 was conducted as a control, without ultrasound. The solution was subjected to the ultrasound throughout the experiment, without any seed, but with mechanical agitation, so as to avoid the formation of ultrasonic standing waves. An example of the calculation of constant A is given in Table 2 for the experiment without ultrasound. The results of experiments 2-4 were analyzed in the same fashion.

The calculated approximate At values of the experiments are shown in the figure, in which the continuous line represents the theoretical curve of relationship (1), with the values of $At_{\exp} = f(x/x_N)$ plotted on it.

TABLE 2. The Calculation of Constant A

Condition of experiment	Calculated parameters	1	2	3	4	5	6	7	8	9	10	A_m
Without ultrasound	x/x_N	1	0.945	0.92	0.72	0.67	0.53	0.27	0.25	0.15	0.08	
	t_{exp}, min	0	120	150	240	300	360	480	540	630	600	
	At_{theor}	0	1.15	1.30	2.10	2.25	2.75	3.70	3.85	4.3	5.0	
	A	0	0.0097	0.0087	0.0088	0.0075	0.0083	0.007	0.007	0.0068	0.076	0.008
	$A_m t_{exp}$	0	0.96	1.2	1.9	2.4	2.9	3.8	4.3	0.05	5.3	

We see from the figure that the experimental data agree fairly well with the theoretical curve over the range of relative supersaturations $x/x_N =$ 0.2 to 1 in experiments both with and without ultrasound. For $x/x_N < 0.2$ there is a departure of the experimental points from the theoretical curve.

The data obtained agree completely with the results of Deryabina and Mishchenko, who showed experimentally that the crystallization of gypsum from aqueous solutions of NaCl and MgCl$_2$ without ultrasound obeyed relationship (1) for relative-saturation values between 1 and 0.2, while for $x/x_N < 0.2$ there was also a deviation of experimental points from the theoretical curve [4].

Thus the kinetics of the decomposition of supersaturated aluminum fluoride solutions with and without ultrasonic fields of various frequencies and intensities, and without seeds, obey the theroetical Roginskii—Todes equation derived for crystallization without allowing for physical factors such as ultrasound. Hence we may conclude that ultrasound does not affect the mechanism of the decomposition of supersaturated solutions, and constitutes only an accelerating factor, acting in the same way as an increase in temperature or agitation. The deviation of the experimental points from the theoretical curve both with and without ultrasound takes place in all cases for $x/x_N < 0.2$. Apparently this phenomenon is not of a random nature, nor is it the result of measuring errors; it is due to some more fundamental cause which involves peculiarities in the mechanism of decomposition of supersaturated solutions.

The likeliest reason for the deviation of experimental points from theoretical for $x/x_N < 0.2$, in our view, is that the equation does not allow for the twin mechanism of crystal growth: by the formation of two-dimensional nuclei, and through the development of screw dislocations. From the data of Schlipf [5] and Todes [6] it follows that, for considerable supersaturations, the growth of crystals is determined by the probability of forming two-dimensional nuclei on growing faces. With the lowering of the supersaturation below a certain value, the dislocation growth mechanism begins to predominate. Relation (1) is derived on the assumption that the kinetics of the decomposition of supersaturated solutions and the growth of crystals depends only on the probability of forming certain nuclei. Hence it is the failure to allow for the effect of the dislocation growth mechanism which makes the experimental points deviate from the theoretical curve. Analysis of the data presented in the figure confirms this idea. We see from the figure that, for high values of x/x_N, the experimental points lie in the main below the theoretical curve, and for $x/x_N < 0.2$ to 0.3 they are above it. This kind of distribution of experimental points may be explained by the fact that, at high values of x/x_N, the decrease in supersaturation is determined by both the rate of forming nuclei and the development of screw dislocations. At low x/x_N the formation of nuclei practically stops, and the fall in supersaturation is determined only by the development of screw dislocations; this is why the decomposition rate becomes smaller than theoretical.

Since the nature of the experimental point distribution does not change under the influence of high- and low-frequency ultrasound, the ultrasound also has no specific effect on the growth kinetics of AlF$_3 \cdot$ 3.5H$_2$O in stable supersaturated solutions.

Conclusions

1. The kinetics of the decomposition of supersaturated aluminum fluoride solutions in ultrasonic fields obey the theoretical relationship derived for the general case of crystallization from solutions [3].

2. Ultrasound has no effect on the mechanism of the decomposition of stable supersaturated aluminum fluoride solutions and the growth of the crystals.

Literature Cited

1. L. A. Dunaevskaya, S. I. Rempel', and Yu. N. Tyurin, Collection: "Application of Ultrasound in Industry." Vol. 2 (Moscow, 1960).
2. A. P. Kapustin, Effect of Ultrasound on the Kinetics of Crystallization (Izd. Akad. Nauk SSSR, Moscow, 1961).
3. S. Z. Roginskii and O. M. Todes, Izv. Akad. Nauk SSSR, Otd. Khim. Nauk 331 (1940).
4. N. V. Deryabina and K. P. Mishchenko, Probl. Kinetiki i Kataliz, Akad. Nauk SSSR, VII:123 (1949).
5. I. Schlipf, Z. Kristallographie 107(1-2):35-64 (1956).
6. O. M. Todes, Probl. Kinetiki i Kataliz, Akad. Nauk SSSR, VII:100 (1949).

DECOMPOSITION OF ALUMINATE SOLUTIONS
UNDER THE INFLUENCE OF ULTRASOUND
AND WITH MECHANICAL AGITATION

V. A. Derevyankin, V. N. Tikhonov, and S. I. Kuznetsov

It was shown in [1-4] that, on decomposition of aluminate solutions without agitation, both with and without seeds, hydrargillite crystals grow mainly by forming antiskeletal dendrites. Lamellar branches are formed mainly in the pinacoid plane. Prismatic growth is practically absent. The lamellar branches intertwine with each other, and they may very well grow together, or else deform each other in growing. This explains the bulging observed in polycrystalline lamellas of the hydroxide crystallizing out at the bottom and around the walls of the vessel from supersaturated aluminate solutions.

With the agitation of aluminate solutions, especially in the presence of a seed, the pseudohexagonal tabular crystals of hydrargillite are covered with lamellar branches, which may be deformed, and in these dislocations develop. This leads to the appearance of new lamellar branches and the consequent formation of antiskeletal lamellar dendrites of the same structure as in the decomposition of aluminate solutions without agitation.

Sources of the preferential development of dislocations near the center of the growing face of the crystal [5] may include mechanical deformations [6-8], the existence of dislocations in the seed crystal [9], distortions of the dendrite branches as they meet one another [9-11], among others; in the presence of deformations caused by impurities [12, 13], dislocations may develop over the whole volume of the crystal [14, 15]. In the presence of a screw dislocation in a lamellar branch, this may sometimes grow into a prism.

During the decomposition of agitated aluminate solutions (containing a seed), the lamellar and prismatic growths in the antiskeletal lamellar dendrites at first retain the orientation of the parent single-crystal pseudohexagonal nucleus. On further growth of the crystal, under the influence of mechanical deformations [16] and impurities, defects may develop in the crystal lattices of the branches. Individual, widely separated branches will have a certain deviation in the orientation of their faces with respect to each other. * The configuration of the branches may also change owing to the nonuniform access of the supersaturated solution to them. Thus, as the crystal grows, mosaic structure develops. In individual cases, twinning deformation appears, since the conditions facilitating the development of an ordinary dendrite are extremely favorable to the formation of twin dendrites (growth twins). The deformation of the dendrites causes the number of growth directions to increase, and the dendrites to lose their single-crystal structure, transforming into crystals of comparatively equiaxial form, somewhat similar to spherulites. Such comparatively equiaxial crystals of hydrargillite are called "pseudospherulites."

Thus the agitation of aluminate solutions in the course of their decomposition substantially changes the habit of the hydrargillite crystals being formed. Moreover, it is known that the decomposition rate of solution rises considerably if these are agitated by means of ultrasound of a certain frequency and power.

When the crystallizing medium is treated with ultrasound, the crystallization rate may rise because the probability of nuclei being formed spontaneously is increased [17], the role of frictional forces [18] between the liquid phase and the growing crystals rises (breaking these up and dispersing them [17-19]), the steady vibra-

---

*This lack of parallel disposition between the branches of dendrites is found very often in nature [16].

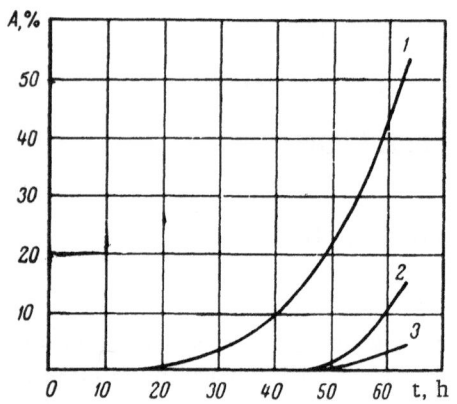

Fig. 1. Decomposition rate of aluminate solutions. Means of agitation: 1) by ultrasound; 2) by magnetic stirrer; 3) none.

tions of cavitation bubbles facilitate the dispersion of the crystals [18], and the role of insoluble impurities as seeds increases [17]. All this leads to the formation of a large number of crystallization centers.

The linear growth rate of the crystals rises under the influence of ultrasound only in weakly supersaturated solutions and for low ultrasound intensities. For large intensities, dispersion of the crystals occurs [19].

It is pointed out in [18] that intensive crystallization of a melt first takes place near the source of ultrasonic oscillations, and then extends over the melt in the form of individual regions.

Kapustin [19] observed an avalanche-like crystallization of salts from solutions under the action of ultrasound.

The mechanism of the effect of ultrasound on the crystallization rate of hydrargillite, and also the ways in which its particles grow, has unfortunately been poorly studied. The aim of this paper is to enhance our understanding of this subject as far as possible.

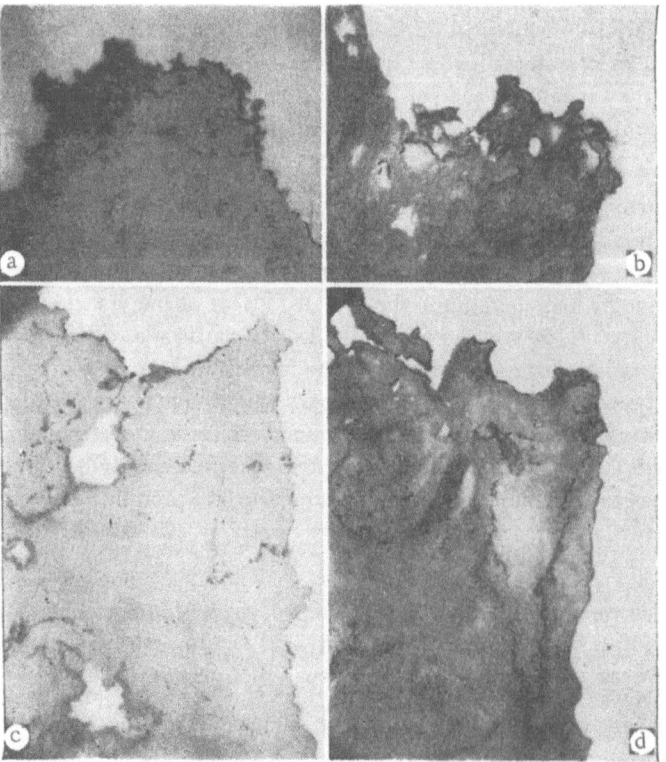

Fig. 2. Microphotographs of aluminum hydroxide crystals obtained on the decomposition of aluminate solutions without agitation. Temperature 36°. × 29.

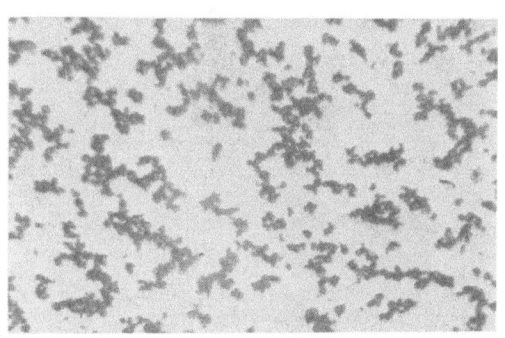

Fig. 3. Microphotographs of aluminum hydroxide crystals obtained on the decomposition of aluminate solutions with agitation in the magnetic stirrer. Temperature 36°. × 70.

The original aluminate solution for the experiments was obtained by dissolving aluminum of the A_{00} type in a solution of NaOH; it had the following composition: 131.8 g/liter Al_2O_3 and 136.5 g/liter Na_2O (molecular ratio $Na_2O : Al_2O_3 = 1.7$). The decomposition of the solution was effected both with and without an aluminum hydroxide seed, obtained from the Bogoslovskii Aluminum Works. The seed ratio (ratio of the Al_2O_3 in the seed to that in the solution) was 1.0.

The decomposition of the solution without a seed was effected in glass beakers and that with the seed in glass retorts hermetically closed by rubber bungs previously boiled in alkali. The beakers were placed in an ultrathermostat of the E-149 type or an electromagnetic stirrer of the MM-2 type with heating, and the retorts in an air-agitation thermostat. The agitation of the pulp in the latter case was effected by rotating the retort "head-over-heels" at a rate of 26 rpm.

The aluminate solutions without seeds were agitated by an electromagnetic stirrer or subjected to ultrasound at 830 kc/sec and 2 W/cm^2, and also held without agitation at a fixed temperature. The source of ultrasound was an ultrasonic therapeutic portable apparatus UTP-1. In decomposing the solutions without seeds, the vibrator was introduced into the solution from above. In the case of decomposition with a seed, the latter was previously subjected to ultrasound for 90 min in water and in the aluminate solution. In some experiments, the ultrasonic field was directed at the hydroxide from above, through the layer of liquid, and in others the vibrator was at the bottom of the vessel in which the irradiation was taking place, and the hydroxide lay directly on the vibrator. The hydroxide treated with ultrasound was used as a seed. For purposes of comparison, parallel experiments with seeds untreated by ultrasound were carried out.

In the course of decomposition, the temperature of the aluminate solution in the beakers was held at $36 \pm 0.5°C$, and for decomposition with a seed (in the retorts) it was gradually lowered from 60 to 46.5°C.

Selection of samples was made after 12, 24, 48, 60, and 68 h. In all cases the precipitates were separated from the mother liquor by filtration. The solutions were analyzed by the method of Bogolepov [20], and the precipitates were carefully washed in a hot-water filter, dried to the air-dry state, and subjected to X-ray structural, phase, and crystallo-optic analysis.

Figure 1 shows the kinetics of the decomposition of aluminate solutions without a seed. As we see, in 68 h the solution decomposes by 5% (3) without agitation, by 17% on mechanical agitation in the magnetic stirrer (2), and by 58% under the influence of ultrasound (1).

The appearance of the first crystallites on decomposition of the aluminate solution without agitation was noted at the walls of the beaker, near the surface of the solution. Upon agitation of the solution with the stirrer, the first crystals were noted in the volume of the solution. The same was found on agitating with ultrasound. In the latter case, marked motion of the solution (macromixing) was not observed. Together with the abundant appearance of suspended matter in the solution, a hydroxide precipitate appeared at the peripheral regions of the vibrator surface.

On decomposition of the solution without agitation, fine, branched lamellar crystals developed (Fig. 2a). At first these grew in the pinacoid plane. Then, as growth proceeded, collisions of the branches (Fig. 2b, c) and deformations arose, and the plate began to warp, developing folds (Fig. 2d). The thickness of the plate, however, hardly increased at all. These lamellar formations collapsed relatively easily.

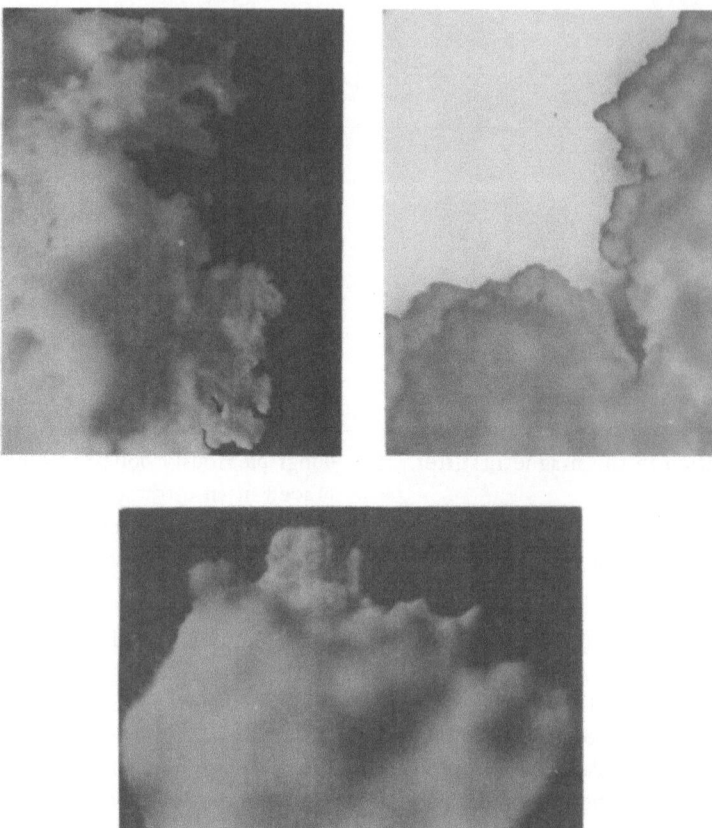

Fig. 4. Microphotographs of aluminum hydroxide crystals obtained
on the decomposition of aluminate solutions under the influence of
elastic oscillations in an ultrasonic field. Temperature 36°. x 29.

When the aluminate solutions were agitated with the magnetic stirrer, finely dispersed precipitates of aluminum hydroxide were obtained (Fig. 3). According to electron microscopic observation [21], the precipitates consisted of tabular or prismatic crystals and sometimes comparatively equiaxial particles of hydroargillite.

In an ultrasonic field, under the influence of elastic oscillations, severely deformed lamellar dendrites up to 15 or 20 mm thick were formed. The growth of these dendrites took place as an avalanche (Fig. 4). The crystallization rate of the hydrargillite in the ultrasonic field was considerably higher than for ordinary mechanical agitation.

The hydroxide precipitated under the influence of ultrasound was distinguished by reduced mechanical strength.

The experimental data obtained suggest that, on the decomposition of solutions under the influence of ultrasound, an important part is played by frictional forces at the melt/crystal surface, these not breaking the growing crystal, but straining it, creating numerous defects, such as dislocations, in the crystal lattice. Such deformations are especially possible in thin branches. Here, apparently, polymerization of the aluminate ions [22] takes place more intensely:

$$m\,Al(OH)_4^- \longrightarrow Al_m(OH)_{4m}^{m-}.$$

The development of defects on the surface of growing crystals and the polymerization of the solution are evidently also causes of the accelerated growth of hydrargillite crystals. In the initial period of decomposition of the solution, polymerization plays the main role. With the appearance of the solid phase, the leading role shifts to defects on the crystal surface.

We find that treatment of the seed with ultrasound before introducing it into the aluminate solution facilitates deeper decomposition of the latter. No substantial change occurs in the coarseness of the seed crystals in this. The coarseness of the aluminum hydroxide forming during the decomposition, however, depends considerably on the intensity of the irradiation. Thus, for example, on introducing the vibrator from above the solution, i.e., on treating the seed in a weaker field, a three-times-smaller $10-\mu$ fraction was obtained in the resultant hydroxide than in the control sample. In experiments in which the seed was treated directly on the surface of the vibrator, the hydroxide produced had a $10-\mu$ fraction 1.5 to 2 times larger than that of the control.

Thus the experimental data enable us to assert that the accelerating action of elastic oscillations from a weak ultrasonic field during the decomposition of aluminate solutions in the presence of a seed is mainly a matter of deforming the growing crystals, engendering a greater surface activity on these. This further confirms the conclusion of [19] that only a small intensity of ultrasonic field exerts an effective influence on the linear crystallization rate.

Literature Cited

1. V. A. Derevyankin, S. I. Kuznetsov, and O. K. Shabalina, Transactions of the S. M. Kirov Ural Polytechnic Institute. Collection 98, p. 106 (1960).
2. S. I. Kuznetsov, V. A. Derevyankin, and O. K. Shabalina, "Chemistry and Technology of Alumina," (Transactions of the All-Union Conference, Alma-Ata, October 6-9, 1959) (1961), p. 15.
3. S. I. Kuznetsov, V. A. Derevyankin, and O. K. Shabalina, Study of the Dissolution and Growth of Aluminum Hydroxide Crystals in Alkaline Aluminate Solutions, Certificate of Registration of Priority from 22.03.61, No. 22,500, April 10, 1961.
4. V. A. Derevyankin and S. I. Kuznetsov, Tsvetnye Metally (5):46 (1961).
5. F. C. Frank, Disc. Faraday Soc. (Crystal Growth) (5):(1949).
6. F. C. Frank, Phil. Mag. 42:1014 (1951).
7. A. Korndorffer, H. Rahbek, and F. Sultan, Phil. Mag. 43:1301 (1952).
8. A. J. Forty, Direct Observations of Dislocations in Crystals [Russian translation] (Metallurgizdat, 1956).
9. F. C. Frank, Advances in Physics (Phil. Mag. Suppl.) I(1):91 (1952).
10. G. G. Lemmlein and E. D. Dukova, Kristallografiya 1(3):351 (1956).
11. G. G. Lemmlein, E. D. Dukova, and A. A. Chernov, Kristallografiya 2(3):428 (1957).
12. G. G. Lemmlein, Vestn. Akad. Nauk SSSR 4:119 (1945).
13. G. G. Lemmlein, Dokl. Akad. Nauk SSSR, 84(6):1167 (1952).
14. I. M. Hedges and J. W. Mitchell, Phil. Mag. 44:223 (1953).
15. A. Verma, Growth of Crystals and Dislocations [Russian translation] (IL, 1958), pp. 152-53.
16. H. Buckley, Growth of Crystals [Russian translation] (IL, 1954), p. 342.
17. O. V. Abramov, V. E. Neimark, and I. I. Teumin, All-Union Conference on the Theory of Crystal Growth and Phase Transformations, Summaries of Contributions (Izd. Akad. Nauk Belorus. SSR, Minsk, 1961), p. 23.
18. I. G. Polotskii and G. I. Levin, All-Union Conference on the Theory of Crystal Growth and Phase Transformations, Summaries of Contributions (Izd. Akad. Nauk Belorus. SSR, Minsk, 1961), p. 22.
19. A. P. Kapustin, All-Union Conference on the Theory of Crystal Growth and Phase Transformations, Summaries of Contributions (Izd. Akad. Nauk Belorus. SSR, Minsk, 1961), p. 22.
20. N. I. Bogolepov, Legkie Metally (4):22 (1936).
21. V. A. Derevyankin, Study of the Processes of Leaching and Decomposition in the Production of Alumina by the Bayer Method (An Examination of the Dissolution and Growth of Aluminum Hydroxide Crystals), Dissertation (Sverdlovsk, 1960).
22. S. I. Kuznetsov, Production of Alumina (Questions of Physical Chemistry) (Metallurgizdat, 1956).

THE EFFECT OF AN ELECTRIC FIELD ON THE CRYSTALLIZATION
PARAMETERS OF A SUBSTANCE

L. T. Prishchepa

Frenkel''s theory of heterophase fluctuations [1] most fully describes the kinetics of transformation phases, and in particular crystallization from metastable, supercooled melts. This theory rests on the general principles of statistical mechanics, which indicate that even in a thermodynamically stable system there must be fluctuations, i.e., local and transitional deviations from the normal state. If these density fluctuations are smallish and lie within limits compatible with the maintenance of a given aggregate state, then these ordinary density fluctuations are called "homophase."

As Frenkel' showed, together with these one must bear in mind density fluctuations which go beyond the limits compatible with the original aggregate state, i.e., which correspond to the formation of nuclei of a new phase, which is in our case crystalline. Such fluctuations have become known as "heterophase."

If the main phase remains thermodynamically stable, the new-phase nuclei arising in it are nonviable, i.e., they develop, reach feeble dimensions, and then vanish, showing no tendency to unlimited growth.

When the main phase is thermodynamically unstable, this tendency becomes dominant in the new-phase nuclei after they have reached certain critical dimensions.

The formation of a crystalline phase in the melt passes through an intermediate state of prenuclear groups, which do not possess surfaces of separation and which are devoid of phase characteristics. The composition of the prenuclear groups is analogous to that of the crystalline phase capable of precipitating from the melt according to the phase rule. The prenuclear groups possess considerable mutual mobility of particles and maintain an excess of free energy relative to the energy level of the nucleus and the melt. If this component of free energy is liberated, there is a change in the character of the bonds in the group, and these pass preferentially out of the associative state into the coordination−chemical state, which also determines the development of a nucleus in the form of a crystal lattice bearing appropriate parameters. Hence the probability of the growth of a nucleus will be determined by those active complexes which are capable of movement, i.e., it depends on the mechanism of the exchange of molecules between the liquid phase and the nucleus. The number of nuclei, as Frenkel' showed, depends exponentially on the activation energy. It follows that the rate of nucleation depends mainly on the activation energy and the work of formation of three-dimensional nuclei, into which enters the interphase surface energy at the separation boundary between solid and liquid phases.

The theory gives the following expression for the rate of nucleation:

$$I = Ce^{-\frac{U}{RT}}\, e^{-\frac{A}{T\,(T_0-T)^2}},$$

(1)

where I is the number of crystallization centers developing per unit volume per unit time, C is a factor independent of temperature, U is the activation energy on transfer of an atom from the crystal lattice of the original phase to the lattice of the nucleus, A is the work of formation of a three-dimensional nucleus, and R is the universal gas constant.

Calculation shows that the work of formation of a three-dimensional nucleus should equal

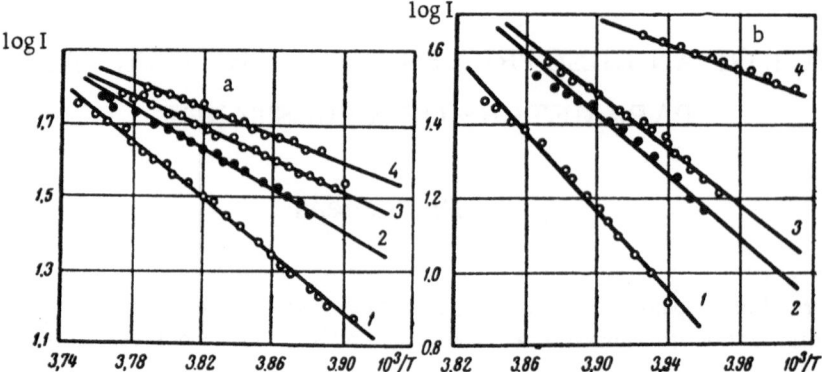

Fig. 1. Variation of log I with 1/T for a) betol and b) antipyrine: 1) control; 2) at frequency 10,000 cps; 3) 940 cps; 4) 2650 cps.

$$A = \frac{16\,\pi\sigma^3 T_0^2 v_B^2}{3\,\lambda^2\,k} \, ,$$

(2)

where σ is the interphase surface energy at the crystal/melt boundary, T_0 = melting point, T = exposure temperature, λ = heat of fusion calculated for one molecule, k = Boltzmann's constant, and v_B = mean volume pertaining to a single molecule.

If we consider the formation of a crystalline phase in the melt under the influence of an electric field, then the latter changes the work of formation of the crystal nucleus, and hence must change the probability of formation, other conditions being equal.

The aim of the present paper is to discover the effect of the electric field on the activation energy and also on the work of formation of crystalline nuclei, or on the interphase surface energy at the solid—liquid surface of separation.

The results of studying the effect of alternating electric fields on the nucleation rate of antipyrine and betol in the frequency range 0 to 10,000 cps are presented in [2]. Experiments by Gorskii and Mikhnevich [3] made with pipeiine and those of Kondoguri [4] made with sulfur confirmed that the electric field had an effect which was manifested as a shift in the temperature/nucleation-rate curves at large supercoolings. Our own preliminary experiments on the effect of a steady electric field on the crystallization of betol and antipyrine, however, gave a negative result; the reason for this lay in the high electrical conductivity of betol and antipyrine resulting from the polarization phenomenon. The field applied to the laminated condenser concentrates in the glass plates. The large electrical conductivity of the substances under examination prevents the effect of the electric field from being observed, and hence an alternating field has to be used in order to eliminate the polarization effect. Under the influence of the alternating electric field, there is a shift in the temperature/nucleation-rate curve to the low-temperature side (large supercoolings). There is a small shift for a 50 cps field, a larger one for 940 cps, and a still larger one for 2650 cps. Upon further raising of the frequency, however, the shift diminishes.

The experimental temperature/nucleation-rate curves constitute the superposition of two curves: one relating to crystallization centers formed in the two layers adjacent to the walls, and the other to centers formed beyond the sphere of influence of the glass surface. The falling part of the resulting curve at high temperatures is due to processes taking place in the volume, and the falling part at low temperatures to processes in the surface layers.

TABLE 1. Variation of Activation Energy
With Frequency of the Electric Field Perpendicular
to the Layer of Material

Frequency, cps	Activation energy, cal/mole	
	Betol	Antipyrine
0	17047	24704
50	13665	20540
940	10860	17739
2650	8419	8363
5600	10450	8552
10000	13462	19515

In the low-temperature part of the nucleation-rate curve, the factor

$$e^{-\dfrac{A}{T\,(T_0-T)^2}}$$

in formula (1) containing the work of formation of three-dimensional nuclei does not have to be considered. It affects the nucleation-rate/temperature curve so slightly that its effect may be neglected at low temperatures [5, 6]. This confirms the fact that the second term of Eq. (1), associated with the mobility of the particles, plays the dominant role in the course of the nuclea-tion-rate/temperature curve. Hence in the small-supercooling region the number of observable crystal nuclei should satisfy the equation

$$I = C e^{-\frac{U}{RT}},$$

(3)

whence

$$\ln I = \ln C - \frac{U}{RT},$$

(4)

that is, the graph relating ln I to 1/T should be a straight line. The slope in coordinates lg I, $10^3/T$ determines 0.4343 U/R, from which the activation energy can be calculated. Figure 1a shows graphs relating lg I to $10^3/T$ from the experimental data from which the temperature curves of the number of crystallization centers described in [2] were constructed. The electric field of various frequencies (calculated strength 4000 V/cm) was perpendicular to the layer of betol (graphs for 50 and 5600 cps omitted).

As seen from Fig. 1, the points fall neatly on straight lines. The resulting values of activation energy for betol and antipyrine appear in Table 1.

In the high-temperature region (small supercoolings) of the nucleation-rate curve, the transformation rate is determined mainly by the work of formation of the nuclei. In this region, the relation between the number of crystallization centers and temperature is different [5] from that in the low-temperature region, and is determined not by their growth on exposure but by formation as a result of fluctuations, so that

$$\lg I = C_2 - \frac{A}{T\,(T_0-T)^2} \cdot 0.4343.$$

(5)

If we set out the values of $1/T(T_0 - T)^2$ along the axis of abscissas and lg I along the axis of ordinates, we obtain a straight line. Figure 2a shows experimental points of lg I as a function of $10^7/T(T_0 - T)^2$ for betol in electric fields of various frequencies (calculated strength 4000 V/cm) perpendicular to a betol layer (graphs for 50 and 5600 cps omitted).

The slope of the straight lines gives parameter A, from which the interphase surface energy at the crystal/melt boundary is calculated. It must be noted that this calculation is approximate, since the values of interphase energy σ, melting point T_0, and heat of fusion λ relate to microcrystalline nuclei, while for the calculation of the value of σ we use values of T_0 and λ determined in experiments with macrocrystals.

If along the axis of ordinates we replace lg I by lg I + 0.4343 U/R, as discussed by Vykhovskii [7], the slope of the straight lines will be approximately the same, as indicated in [8]. Hence we consider that, within

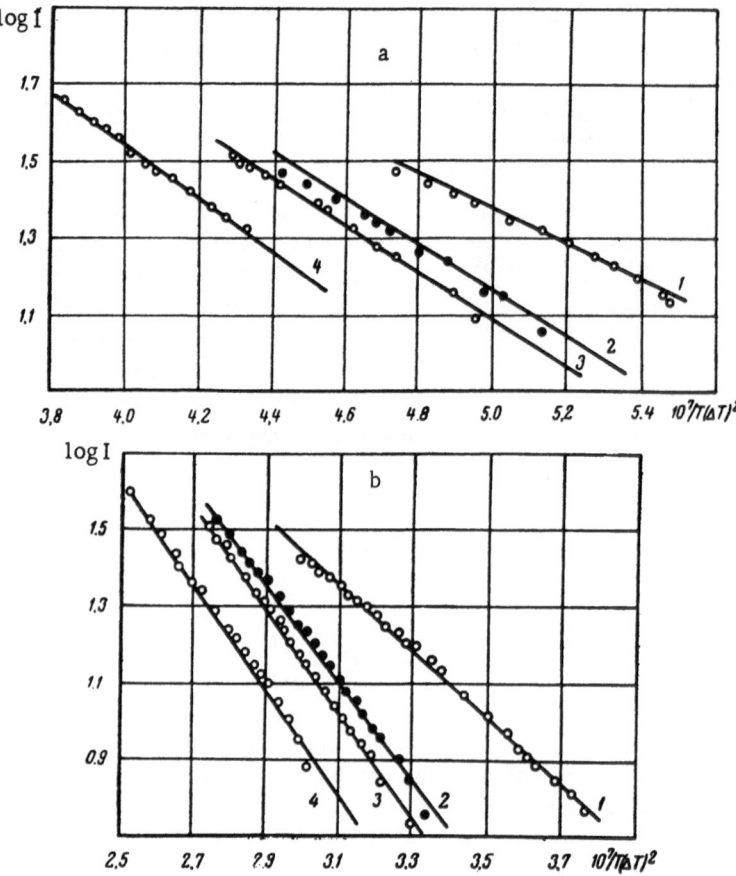

Fig. 2. Variation of lg I with $1/T(\Delta T)^2$ for a) betol and b) antipyrine:
1) control; 2) at frequency 10,000 cps; 3) 940 cps; 4) 2650 cps.

the limits of experimental error, the values of the interphase surface energy are calculated correctly. This method would only lead to an incorrect estimate of the pre-exponential factor in formula (1).

The values of interphase surface energy calculated from the slopes of the straight lines (see Fig. 2), i.e., from the valuesof parameter A, for beol and antipyrine, are shown in Table 2.

We see from Table 1 that, with increasing frequency of the electric field, the activation energy diminishes, and·this in turn leads to an increase in the number of crystallization centers. Analyzing a wide range of experimental material (concerning the influence of various physical factors on the crystallization parameters of matter) on the basis of Frenkel''s fluctuation theory, Gorskii [9] showed that every physical action B, the value of which does not remain constant ($dB/dt \neq 0$), will reduce the activation energy and hence also increase the probability of forming crystallization centers. On the other hand, physical actions constant in time increase the activation energy and have the opposite effect on the crystallization process. In the first case, external physical action increases the heterophase fluctuations in metastable systems and thus accelerates the diffusion of ions in the liquid phase; in the second case, it reduces these and exerts a stabilizing influence on the systems.

In an electric field, polar molecules are acted on by forces tending to turn them so that they fall with their long axes directed in a definite fashion relative to the lines of force, either parallel or perpendicular to these [10, 11]. In the presence of a strong bond between neighboring molecules, their orientation in the field is limited by this bond, and they only rotate through small angles in the external electric field. This relates not only to individual molecules but also to molecular groups.

TABLE 2. Variation of Interphase Energy with Frequency of Electric Field
Perpendicular to the Layer of Material

Frequency, cps	Interphase surface energy, ergs/cm^2	
	Betol	Antipyrine
0	9.048	5.776
50	9.137	6.215
940	9.262	6.686
2650	9.585	6.730
5600	9.374	6.692
10000	9.156	6.589

The parameters of the electric fields (their strength and frequency) have a large influence on processes of orientation polarization. On the effect of the field strength there is a common view that, with increasing strength, the degree of perfection of orientation polarization rises, and for certain values of field strength reaches a maximum, at which a "saturation" phenomenon sets in. If we consider the amplitude of the oscillation of particles around a mean position, then this rises with increasing field strength and decreasing frequency.

Gorskii and Mikhnevich [3] also observed that, for a constant field strength, its effective action was connected with the degree of supercooling of the specimen. We also observed an analogous phenomenon. This is confirmed by the fact that the number of crystallization centers formed under the influence of the field is inversely proportional to the temperature.

On the effect of the field frequency, as shown by experiment, the following treatment will be valid. With increasing frequency, the dipole moments are oriented in the direction of the field. The maximum orientation for betol and antipyrine molecules occurs at a frequency of 2650 cps. At this moment the dipole molecules tend to return to their original state, and the already-formed structures correspondingly break up. In association with this, the activation energy of betol also decreases by 50.6% at a field frequency of 2650 cps, and that of antipyrine decreases by 66.2%. The decline in the activation energy of antipyrine is greater than that of betol because the antipyrine molecule has a greater dipole moment. At frequencies greater than 2650 cps, relaxation of the rotational oscillations sets in, and the molecules are unable to rotate behind the field; the orientation effect diminishes, and, on reaching a frequency of 10,000 cps, the activation energy approaches its value in the absence of field.

We see from Table 2 that, with increasing field frequency, the interphase surface energy rises up to a certain frequency, after which it falls and approaches the no-field value. The maximum interphase surface energy occurs for a field frequency of 2650 cps. At this frequency the interphase surface energy of betol is increased by 5.9%, and that of antipyrine by 16.5%. The change in the interphase surface energy depends on the dipole moments of the molecules, just as does the activation energy.

The interphase surface energy is a quantity which indicates the nature of the interaction forces of the boundary layer between the liquid and the crystal. If it increases, then there is a greater difference between the character of the interaction at the boundary with the solid and phases.

The surface layer of the material is deformed under the influence of the electric field; the number of molecules in the nucleus changes, as does the number of nuclei in the electric field. Hence the interphase surface energy, which depends on the dimensions of the nuclei, will also change. While the molecules can still rotate in the electric field, the interphase surface energy increases. At frequencies greater than 2650 cps, the molecules can no longer rotate along with the field; the interphase surface energy diminishes, and, on reaching a frequency of 10,000 cps, approximates to the no-field value.

The kinetics of the crystallization of supercooled liquids is determined by the interphase surface energy at the solid/liquid separation boundary, as well as by the activation energy.

The experimental material presented above shows that the effect of the electric field on the nucleation rate is obtained by determining the effect of the electric field on the activation energy and the interphase surface energy of the substance in the fluctuation-theory formula (1).

In conclusion, the author takes this opportunity of thanking F. K. Gorskii for his aid in the supervision of this investigation.

Literature Cited

1. Ya. I. Frenkel', Selected Works, Vol. 3: Kinetic Theory of Liquids (Moscow—Leningrad, 1959).
2. F. K. Gorskii and L. T. Prishchepa, Crystallization and Phase Transformations (Izd. Akad. Nauk Belorus. SSR, Minsk, 1962), pp. 386-91.
3. F. K. Gorskii and G. L. Mikhnevich, Zh. Eksp. i Teor. Fiz. 2:264 (1932).
4. W. W. Kondoguri, Phys. Zeits. d. Sowjetunion 9:603 (1936).
5. G. L. Mikhnevich, Dissertation, V. I. Lenin Belorussian State University (Minsk, 1960).
6. V. I. Malin, Zh. Fiz.Khim. XXVIII(11):(1954).
7. A. I. Bykhovskii, Scientific Transactions of the Ukrainian Agricultural Academy, Kiev, IX:431-39 (1957).
8. V. I. Danilov, Structure and Crystallization of Liquid (Kiev, 1956), 407-22.
9. F. K. Gorskii, Effect of Various Physical Factors on the Crystallization Parameters of Matter. Collection of Scientific Works of Minsk Medical Institute, Vol. 15 (1955).
10. V. Frederiks and V. Tsvetkov, Uch. Zap. Leningr. Gos. Univ., Ser. Fiz. Nauk 8(2):3-33 (1936).
11. N. N. Mirolyubov, Zh. Rossisk. Fiz. Khim. Obshch., Ch. Fiz., 57(5-6):435-55 (1925).

EFFECT OF A MAGNETIC FIELD ON THE FORMATION OF CRYSTALLINE NUCLEI IN SUPERCOOLED BETOL

F. K. Gorskii and A. V. Akhromova

The effect of a magnetic field on crystallization processes is of practical interest; it is utilized in industry, for example, in the form of the so-called magnetic treatment of water with the object of preventing scale-formation in steam boilers [1]. There is not much published information in this field, however, and the mechanisms explaining the action of the magnetic field are insufficiently clear.

For the further development of the theory, both new investigations and a critical review of existing material are required.

The first experiments on the effect of a magnetic field on the crystallization of supercooled melts were performed by Kondoguri [2]. Kondoguri studied the time-dependence of the number of crystallization centers in supercooled piperine at room temperature, and established that a magnetic field of 9000 to 10,000 G accelerated the formation of crystalline nuclei. For materials with a high crystallization rate, the method of counting the centers was unsuitable, and Berlaga [3] used another method. The molten specimen was cooled slowly, and the temperature at which crystallization of the specimen took place formed an arbitrary characteristic of the crystallization process with or without the field. After three or four melts of the specimen, stabilization of the crystallization temperature set in, and experiments were made in turn with the magnetic field applied. Both for paramagnetic azobenzene and diamagnetic diphenylamine, the field increased the crystallization temperature, i.e., it acted in the accelerating sense on the formation of nuclei. The curve relating the number of crystallization centers to the temperature, as we know, passes through a maximum. For piperine this maximum is 40°C [Tamman] and for azobenzene −23°C [Tsedrik (4)]. Thus the difference between the experiments of Kondoguri and Berlaga lies not only in the fact that they were made for substances with different linear crystallization rates, but also as regards the temperatures studied. Kondoguri's experiments relate to the low-temperature part of the curve giving the number of crystallization centers, and Berlaga's to the high-temperature part. As indicated above, in both temperature ranges the magnetic field accelerated the formation of crystal nuclei. Gorskii [5] studied the action of a magnetic field over a wide temperature range for piperine. It transpired from these experiments that, in contrast to the action of an electric field, which shifts the temperature curve into the region of high supercoolings, the magnetic field increases the probability of nucleation at all temperatures. The results of this work agree with both the previous workers and interconnect them. Later experiments of Kondoguri [6] on the crystallization of supercooled sulfur in electric and magnetic fields showed that the magnetic field produced a shift in the n.c.c. (number of crystallization centers) curve. Finally, as shown by Mikhnevich and his pupils Zaremba and Efimova [7, 8], the existence of a boundary layer in thin specimens results in a difference between the actions of pulsed (created for a short period) and steady magnetic fields. In the works mentioned, a constant field at 2°C reduced the slope of the first part of the kinetic curve, while a pulsed field increased it. The authors associated the peculiar action of the pulsed field with the collapse of the boundary layer. A steady field strengthens the boundary layer, while a pulsed field destroys it.

Experimental Part

We chose betol as our material for investigation. It was purified by recrystallization from alcohol and filtration through glass filter No. 4. U-shaped glass thermometer capillaries of internal diameter 0.3 mm, previously washed in chromic mixture, distilled water, and absolute alcohol, and well-dried, were filled with

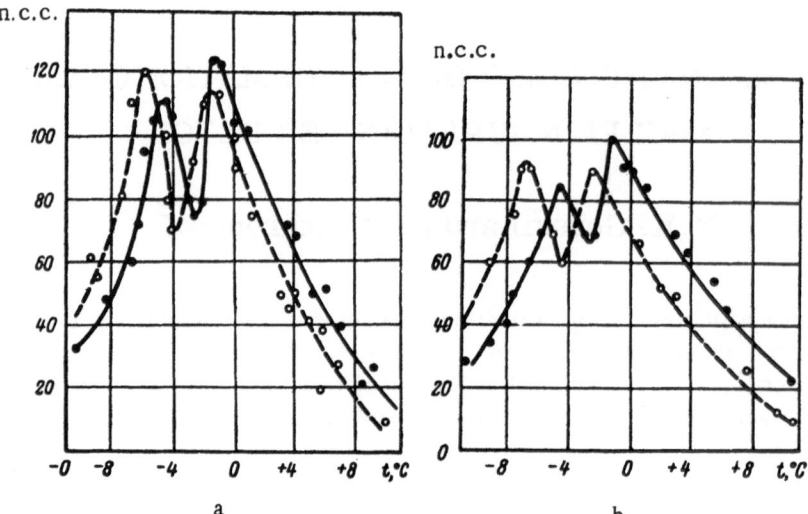

Variation of the number of crystallization centers of supercooled betol
with temperature; continuous line: without magnetic field (control);
broken line: in magnetic field. a) 10,000 G; b) 19,000 G.

molten betol. Between the poles of an electromagnet a miniature thermostat was set for the specimen; this
was connected to an ultrathermostat and cold reservoir. As thermostat liquid we used silicone liquid No. 3.
The use of solid CO_2 and methyl alcohol as contact liquid in the cold reservoir enabled us to hold the temperature in ranges from −20 to +20°C. Development was carried out at room temperature. Experiments were made
at two magnetic field strengths, 10,000 and 19,000 G. In order to eliminate the pulse action of the magnetic
field, a steady current was first switched into the electromagnet from a mercury rectifier, and then the specimen was slowly introduced into the thermostat between the magnet poles. On finishing the exposure, it was
taken slowly out of the field, and only then was the current switched off.

For the pulsed action of the field, the specimen was first introduced into the interpolar space, then the
electromagnetic current was switched on and immediately switched off. The duration of the exposure in one
series of experiments was 3 min and in another 7 min.

Counting nuclei for equal lengths and exposure times enabled us to compare data obtained for different specimens.

In making experiments with successive remeltings of one and the same specimen, it was considered that
reproducible readings were obtained after seven or eight preliminary melts [9]; analogous temperature stabilization occurred in Berlaga's experiments.

Furthermore, for some specimens we began by taking the n.c.c./temperature curve without the magnetic
field and ended by taking it in the magnetic field, while for other specimens experiments with and without the
magnetic field were systematically alternated.

The results obtained are shown in the figure, in a (field 10,000 G) and b (field 19,000 G), from which we
see that the steady magnetic field, just as the case with an electric field, as a result of the orienting action of
the molecules of the melt, shifts the n.c.c./temperature curve in the direction of greater supercoolings.

Only to the right of the maximum does the steady magnetic field reduce the number of centers; to the
left of the maximum this number increases. The volume maximum diminishes slightly, while the surface maximum rises, confirming the view that the boundary layer is strengthened by the steady magnetic field.

The stronger field gives the sharper shift effect (1.2° for 10,000 G and 2.3° for 19,000 G).

Zaremba's conclusion that the steady field reduces the number of crystallization centers for betol is valid only for the high-temperature part of the n.c.c. curve.

In our experiments on the action of a pulsed field on the crystallization of betol, the results entirely coincided with Gorskii's for piperine [6]; this may be explained by the pulsed character of the application of the magnetic field.

Literature Cited

1. V. Yu. Babkina, V. V. Shchetinskii, V. G. Inzhechik, and Ts. L. Drukh, Magnetic Treatment of Water, Tr. NIOKhIM, XI:129 (1958).
2. V. V. Kondoguri, Transactions of the State Physical Institute in Odessa [in Ukrainian], 1(4):(1930).
3. R. Ya. Berlaga, Transactions of the State Physical Institute in Odessa [in Ukrainian], 1(4):(1930).
4. M. S. Tsedrik, Dependence of Polycrystalline Structure on the Temperature of Formation of an Ingot, Dissertation for the Degree of Master of Physicomathematical Science (Minsk, 1957).
5. F. K. Gorskii, Zh. Eksp. i Teor. Fiz. 4(5):522 (1934).
6. V. V. Kondoguri, Transactions of Odessa State University, Physics [in Ukrainian], Vol. II, Odessa, pp. 77-89 (1940).
7. G. L. Mikhnevich, V. G Zaremba, and V. P. Efimova, Kolloidnyi Zhurnal, 24(4):488-90 (1962).
8. G. L. Mikhnevich and V. G. Zaremba, Kolloidnyi Zhurnal, 24(4):491-93 (1962).
9. E. N. Ovchinnikova, Transactions of Odessa University, XVI(72):63-68 (1952).
10. G. V. Zaremba, Transactions of Odessa State University [in Ukrainian], 148:31-35 (1958).